LIVING IN SPACE

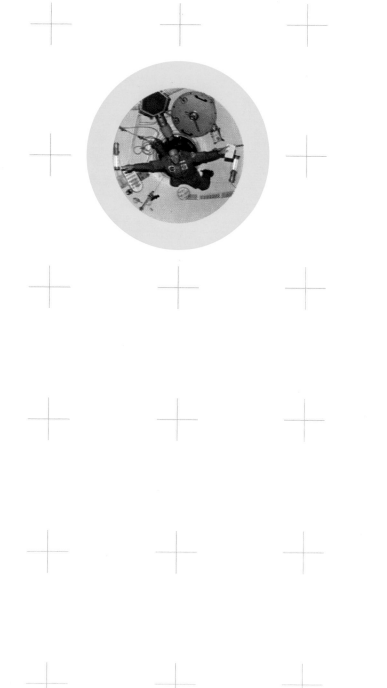

FIREFLY BOOKS

Mondadori would like to join Giovanni Caprara in thanking Alenia Spazio for its invaluable support in the publication of this English version of *Living in Space*. In particular, our gratitude goes to David Climie for the effort spent in translating the original Italian.

A FIREFLY BOOK

Published by Firefly Books Ltd. 2000
First published in Italian as *Abitare lo Spazio* by Arnoldo Mondadori Editore, S.p.A.

First Printing

U.S. Cataloging-in-Publication Data is available.

Canadian Cataloguing in Publication Data

Caprara, Giovanni
   Living in space : from science fiction to the International Space Station

Translation of: Abitare lo spazio.
Includes index.
ISBN 1-55209-549-5

1. Space stations. 2. Manned space flight. 3. Astronautics—International cooperation.
4. International Space Station. I. Title.

TL797.C3613 2000   629.44'2   C00-930873-3

Published in Canada in 2000 by
Firefly Books Ltd.
3680 Victoria Park Avenue
Willowdale, Ontario
M2H 3KI

Published in the United States in 2000 by
Firefly Books (U.S.) Inc.
P.O. Box 1338, Ellicott Station
Buffalo, New York
14205

English text edited by Christine Kulyk

Printed and bound in Spain
D.L. TO: 1329-2000

Contents

For Francesco and Gianluigi as they look toward the future

Foreword

It is difficult to describe the experience of being in space for a long time. It is a complex mix of emotion, new sensations and new perspectives. Even in the company of other crew members there is a sense of solitude and total reliance on the hardware that keeps you up there and on the people on the ground who support you. The foundation of any space mission is the hard work and dedication of the scientific, technical and industrial communities.

It is therefore fitting that Giovanni Caprara's book recounts the history of human adventures in space without omitting any of the people who have contributed to what is an extraordinary achievement—from the early dreamers and pioneers to the groundbreakers and builders. It is also appropriate that Alenia Spazio, one of Europe's major industries for manned space systems, should be a sponsor of the publication. When my friends at Alenia invited me to write this short introduction, I had no hesitation in accepting.

The first thing that struck me about Living in Space *was the wealth of illustrations. Just a cursory glance gives an immediate idea of the excitement of space. But, obviously it is the text that involves us in the trials, tribulations and successes in the history of humankind's attempts to go beyond the atmosphere and stay in the cosmos for longer and longer periods. Now that the International Space Station is real and a home in space for people around the world, this book is a dream come true. Congratulations to Giovanni Caprara, Alenia Spazio and everyone involved in recounting an amazing story.*

Colonel Gerald P. Carr, USMC (Ret.)
Former NASA Astronaut
CDR, Skylab III (SL-4)

★ ★ ★ ★ ★ ★ ★

Preface

It took a century for the dream of a large space station in orbit around the Earth to become a reality. Russian pioneer Konstantin Tsiolkovsky was the first to attempt to give concrete form to the science fictional visions of the late 19th and early 20th centuries. After the initial conquest of space with Sputnik 1, a series of projects, attempts and successes brought the grand idea to maturity, giving rise to the first embryonic space laboratories, from Salyut to Skylab. These were the first, hesitant steps in learning how to build a "cosmic house" where humans could live untroubled and work for the benefit of science and life on Earth.

The odyssey of Russia's Mir closed the pioneering age, opening the way for a large, stable, orbiting settlement. Thus was born the International Space Station, the largest engineering undertaking of all time and the extraordinary result of a common effort that unites the resources, abilities and desires of many nations. In this way, it becomes the first symbol of humankind united in the exploration of space: humanity in space representing planet Earth.

This book recounts a 100-year-long story of human achievement, scientific and technological enterprises, politics and fantastic dreams. In space, imagination and the spirit of adventure are the basic ingredients for every past and future initiative.

I would like to thank Romano Barbera, head of the Program Integration Department at ESA's Directorate of Manned Spaceflight and Microgravity; Elena Grifoni of the Program Integration Department at ESA's Directorate of Manned Spaceflight and Microgravity; Franco Bonacina, head of the ESA press office; Giovanni Rum, head of the ASI space-station program; and Ernesto Vallerani, former president of Alenia Aerospazio, for their reading of certain parts of the book.

I would also like to thank Saverio Lyoi, head of orbital infrastructures at Alenia Aerospazio, for his patient technical explanations.

G.C.

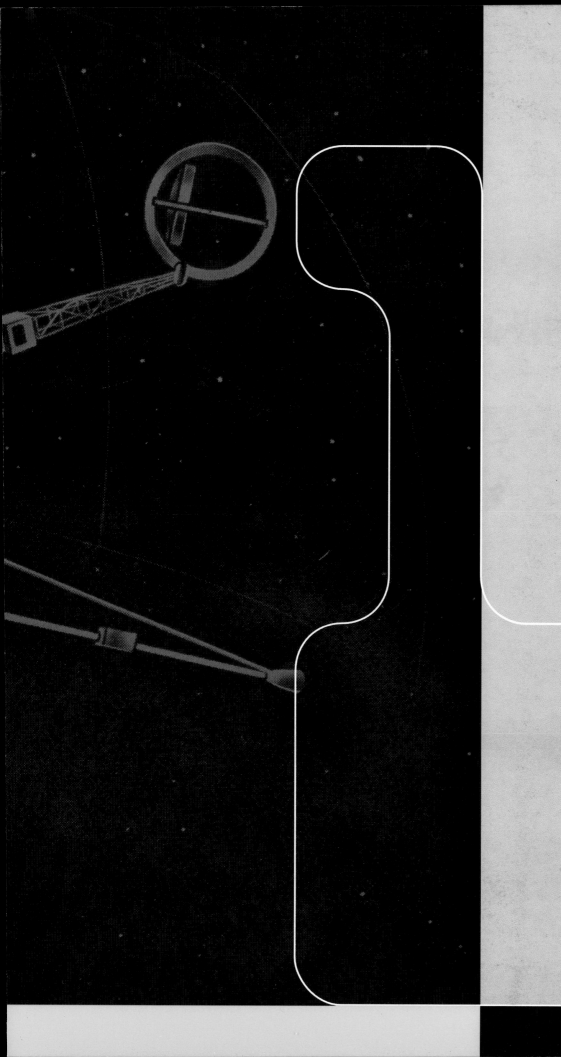

1

Science fiction and pioneer visions

From science
fiction dreams to
von Braun's wheel

Science fiction and pioneer visions

1869: The American Civil War has been over for four years, and it is also four years since the publication of Jules Verne's *From the Earth to the Moon* in France. In Russia, Dmitri Ivanovich Mendeleev is compiling his "periodic table of the chemical elements"; the Suez Canal is opening in Egypt. The same year, the American magazine *Atlantic Monthly* carries a story whose title, "The Brick Moon," betrays an unusual subject. The author is Edward Everett Hale, a writer fascinated by nature and science.

SCIENCE FICTION: "THE BRICK MOON"

The story begins in Boston, with three students remarking that while the heavens have given us the Pole Star to determine latitude, no corresponding reference for longitude exists. To fill this gap, they envision launching a large satellite into a polar orbit. Being visible from the Earth's surface, it would be a valuable aid for navigators.

Thirty years later, the three young men have become Mr. Q. George Orcut, a businessman whose wealth comes from sound investments in railroads, and Messrs. Frederick Ingham and Ben Haliburton, the first a founder of a church and the second of a school. In other words, they are now affluent men who could make their old idea a reality. (Initially, they believe it will cost $60,000, but in the end, it would require $214,729.) So they begin the great adventure on their

lands, which are rich with the raw materials for making bricks and the waterfalls to produce energy.

Their plan is to build a 65-meter-diameter sphere using brick because of its good heat resistance. Placed in orbit at 6,000 kilometers, they explain, the artificial star would be visible from distances over that figure. They discuss several ways to launch the sphere and eventually choose the "two flywheel" technique. This involves two gigantic wheels, the largest ever built. Arranged horizontally and just touching, they are to rotate in opposite directions. The rotation is to be imparted by waterfalls; after some years of continuous turning, the wheels would store up enough energy for the launch.

In the meantime, a construction site is prepared for the sphere at the upper end of a canal. Once completed, the sphere would slide down the canal, terminating at the flywheels. These in turn would impart the thrust needed for the sphere to take off into space. Furthermore, one of the wheels is to have a slightly smaller diameter, and this difference would result in a trajectory curved sufficiently so that the sphere can reach the desired orbit.

Work begins, and the flywheels start to turn. At the same time, a little way off, the sphere begins to take shape at the upper end of the canal. But one night, a problem in the support structure suddenly causes the sphere to slip down the canal, fall onto the wheels and be launched into space.

The three inventors had never planned that their "brick moon" would be inhabited. But what happens changes the destiny of the artificial satellite. When the accident occurs, George Orcut and a group of technicians happen to be inside the sphere with their

Preceding pages: Hermann Oberth's "trampoline station." Below, construction of the "brick moon." On facing page, another illustration of the brick moon from Edward Everett Hale's story.

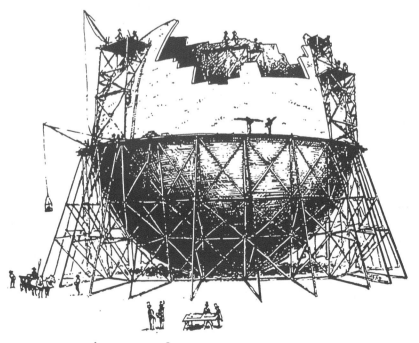

visiting families. And so they find themselves unwitting inhabitants of the space station.

For a year, observers on Earth search for the sphere without success. Then, a Professor Karl Zitta of Breslavia announces in the magazine *Astronomische Nachricten* (Astronomical Information) that he has found it at an altitude of 8,000 kilometers and has named it Phoebe.

Haliburton and Ingham, the two partners left on Earth, step up the search and discover that 37 people are producing a Morse code message by jumping up and down on the sphere in an attempt to communicate with Earth. The message says there is plentiful air and supplies onboard, plus vegetation: palm trees, breadfruit and banana trees. Crops, too, have succeeded, with oats, corn, rice and wheat yielding 10 harvests per year.

To reply to the messages from space, Haliburton and Ingham use huge black letters written on snow. To relieve the solitude of the spacewrecked crew, they decide to send presents, which they wrapped in layers of paper that would burn up in the atmosphere, leaving the contents intact. Some of the presents arrive at their destination, while others are lost, and still others remain in orbit like satellites around the brick moon. In the end, life takes on a normal routine on the sphere, and the inhabitants are content with their incredible destiny.

Edward Everett Hale's story is science fiction's first blueprint for a manned space station. Despite its fantastical nature, it contains many of the fundamental elements necessary for an orbiting base with people onboard. Other stories followed Hale's. In 1897, Kurd Labwitz, a German mathematics teacher, philosopher and science historian, wrote *Auf zwei Planeten* (The Two Planets), which describes a space station orbiting Mars and equipped with an antigravity system. H.G. Wells and Jules Verne also wrote similar stories at the end of the 19th century.

TSIOLKOVSKY AND THE FIRST STATION IDEAS

However, it was not until 1902 that the first pioneers were able to tackle the question more scientifically.

In that year, as the Wright brothers at Kitty Hawk

Konstantin E. Tsiolkovsky
(top) and Robert H. Goddard.

beach in North Carolina were preparing the aeroplane that would make its first historic flight the following year, Konstantin Eduardovich Tsiolkovsky in Russia published his prophetic study *Exploring Universal Expanses With Jet Instruments*. This was later updated in a new edition in 1911-12, in which Tsiolkovsky demonstrated the possibility of using jets to fly into space and presented his first reflections on the potential for space stations. He wrote, "We have the ability to build a permanent observatory moving around the Earth for a long, indefinite time beyond the limits of the atmosphere, just like our Moon."

A scientist of genius, Tsiolkovsky was born in 1857 to a humble family at Izhevsk, in the province of Ryazan. Isolated from the world by deafness resulting from a childhood bout with scarlet fever, he educated himself at home and then studied in Moscow. In 1879, he became qualified as a science teacher. He was assigned to a school not far from the capital, in Borovsk, in the province of Kaluga, and here he spent his life laying the foundations of space flight with his ideas, research and some rudimentary experiments. Through these, he became universally recognized as the "first father of astronautics."

Tsiolkovsky designed rockets and calculated fuel requirements, suggesting that the best mix was hydrogen-oxygen (now used in the engines of NASA's space shuttle). He imagined, for the first time, multistage space vehicles. He also considered the possibility of permanent human settlement beyond Earth, saying that humanity had to learn to leave its planet of origin and to colonize the solar system. He is remembered for his famous comment: "The Earth is the cradle of mankind. But man cannot live in the cradle forever."

In a series of articles published in the Moscow journal *Aviation Reporter*, he laid out various station concepts and discussed their components. The manuscripts, now kept in the Moscow Academy of Science, show that his first general explorations of the subject date from 1878-79. These were then collected and published in 1883 in *Free Space*, a book that dealt with interplanetary voyages. The rings of Saturn inspired him to design a station shaped like a wheel: "An artificial ring allows a person to move freely in all directions," he said.

Tsiolkovsky realized, however, that astronautics was a totally new venture that had never been considered before and whose fascinating possibilities were therefore unknown. With this in mind and to help spread the new concepts, he published technical studies as well as more or less fantastic stories for a wider public in which he explained and anticipated future conquests of the cosmos. He even imagined building a system of space stations around Earth to form a sort of ring that "would receive sufficient energy from the Sun to guarantee life for 20 billion inhabitants." To make life onboard more pleasant, the scientist proposed varying the speed of rotation of the wheel stations to generate the desired amount of gravity, almost replicating terrestrial conditions. Furthermore, he described space stations that would travel around the solar system and orbit other planets to make studying them easier.

The 1912 edition of *Exploring Universal Expanses With Jet Instruments* took his explorations further, and another step forward came with the third edition in 1923. The stations, he wrote, should be located "2,000-3,000 *verst* [a Russian unit of measure; equivalent to about 1,060 meters] away from Earth like moons. Piece by piece, colonies will be formed with materials, machinery and structures brought from Earth. Then, an independent, albeit initially limited, production could gradually develop."

Tsiolkovsky's station design.

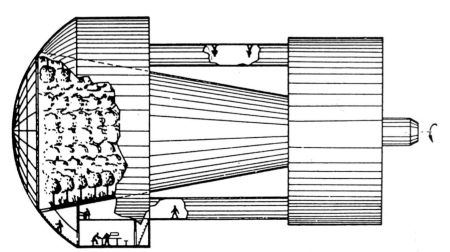

Tsiolkovsky imagined the station as a small planet, divided into two sections, where life would develop independent from Earth. The first section was reserved for passengers. It was less than 400 square meters in area, 100 meters long and divided into connecting cabins. With a shape partly conical and partly cylindrical, it had rocket engines to maintain its desired position and to impart rotation, creating artificial gravity. Above this structure was the second section, with greenhouses 500 meters long by 3 meters wide, containing vegetation to renew the air and where day and night could be lengthened or shortened at will. The plants in the tidy gardens, fertilized by excrement and bacteria, helped compensate for the inhabitants' respiration. The shrubs could be disinfected using solar heat and the vacuum of space.

Perhaps the Russian scientist's most significant idea, however, was that of a self-contained biological cycle onboard. He conceived of transporting organic compost, chemicals, plants, animals and bacteria from Earth to the orbiting bases to create an ecology that would then become self-supporting and completely autonomous, guaranteeing the inhabitants ideal conditions and necessary resources. Everything would be nourished by energy from the Sun. Furthermore, he made the even more ambitious proposal of transforming part of the station into a production center for various things, from food to drugs. He also suggested harvesting asteroids to extract their raw materials.

Even a thrifty proposal (made in 1913) to build a station with used rocket stages in orbit was among Tsiolkovsky's many plans. This idea was revived in the 1980s, when some studies, including NASA's, suggested collecting and reusing hydrogen and oxygen tanks from the shuttle as elements for a space base.

Many of these ideas found their way into film with the 1933 movie *Space Voyage*, for which the Russian scientist was consultant to Mosfilm Studios. By then, his fame was already assured, and the early years he had spent as a researcher with little acceptance were long past. In fact, after the October Revolution, the Kremlin recognized the merits of the great scientist and helped him until he died at his home in Kaluga on September 19, 1935.

FROM RUSSIA TO GERMANY

Unfortunately, because his secluded life prevented scientific exchange and since he published exclusively in Russian—a language little spoken beyond the borders of the Soviet empire—Tsiolkovsky's work remained unknown in the West until the late 1930s, although something of his was translated into German for the first time in late 1920. That same year in the United States, Robert Goddard, the pioneer of liquid-fuel propulsion, suggested using extraterrestrial materials from the Moon and asteroids to build large space bases.

In the same period in Germany, another scientist was exploring the theoretical aspects of long-term settlements in the cosmos. Hermann Oberth, the "second father of space," was born June 25, 1894, in Hermanstadt (then part of Hungary, now in Romania). He trained in Germany, studying at Göttingen, Munich and Heidelberg. In 1923, Oberth published a book containing his doctoral thesis from the preceding year at the University of Heidelberg, where it had been turned down. He obtained a degree in mathematics and physics, however, from the University of

Above, sketch of a space-station design by Tsiolkovsky. Left, drawing of Tsiolkovsky. Below, Tsiolkovsky monument in Moscow.

Hermann Oberth and his "trampoline station" design.

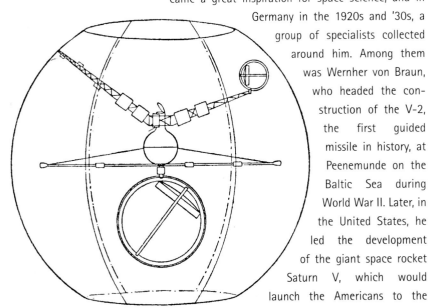

Kalusenburg, in Transylvania, and went on to teach at Schassburg and Medias, where he generally resided until 1938.

Oberth's 1923 book was entitled *Die Rakete zu den Planetenraumen* (The Rocket in Interplanetary Space), and it became a milestone in astronautics. It described, together with the relevant mathematical formulas, the voyage of a rocket into space, the performance of a rocket engine in a vacuum (the subject of much discussion among scientists), the placing of a satellite into orbit and the possible effects of space travel on an astronaut. It also gave details of two types of rocket and explored the concept of a space station. In a second book (*Wege zür raumschiffahrt*) in 1929, Oberth probed more deeply into this latter topic, which he revisited in later years in other works.

While Tsiolkovsky regarded the orbiting base as ideal for the observation of space, Oberth saw it as a valuable platform to observe Earth for a variety of purposes. First among these was science, "as every detail can be seen," but he also noted the military advantages. He foresaw uses for meteorological observation and for telecommunications (initially optical, then radio). Moreover, he considered using the station as a supply base for interplanetary vehicles.

Oberth (who took German citizenship in 1940) became a great inspiration for space science, and in Germany in the 1920s and '30s, a group of specialists collected around him. Among them was Wernher von Braun, who headed the construction of the V-2, the first guided missile in history, at Peenemunde on the Baltic Sea during World War II. Later, in the United States, he led the development of the giant space rocket Saturn V, which would launch the Americans to the Moon.

HERMANN OBERTH'S SYSTEM OF STATIONS

Over 30 years, Oberth designed three types of station, or, rather, a "system of stations," with different functions. The first, called "the trampoline station," or "space port," was to be placed in an orbit close to Earth (300 to 400 kilometers). It would be easy to reach and included an assembly area for vehicles for interplanetary exploration, plus a pear-shaped living area 2,000 meters distant. The sections were held together with a network of cables to which protective bombs were attached. The station was also to have growing plants to generate oxygen.

The second type of station was named "sidereal day," because it was on a higher orbit (35,700 kilometers) that would circle the planet at the same speed as the Earth's rotation and therefore would not appear to be moving over the course of a day. This station would have special uses—for example, as an astronomical observatory—and its structure would be adapted to the job at hand.

The third station, called "strategic," would meet military needs, located in a relatively low orbit (approximately 600 kilometers) but heavily inclined with respect to the Earth's equator. "The best solution," wrote Oberth, "is to make it perpendicular [to the equator] so that the orbital plane and that of the equator form a 90° angle. In this way, it would be possible to attack any point on the Earth's surface at least twice each day."

Oberth's interest then turned to an unmanned version of the station. This was a "space mirror," which he first described in 1923. The mirror was 100 to 150 kilometers in diameter and would be assembled by astronauts in space. The basic element was a metallic mesh with myriad small, circular, movable mirrors. Oberth calculated that an orbiting reflector like this, placed at an altitude of 7,700 kilometers and radiating perpendicularly, would illuminate an area of 5,500 square kilometers of the Earth's surface with an intensity no more than that found naturally at the equator. "With individual surfaces of the mirror," the scientist specified, "large cities can be lit up at night; and a greater concentration could melt dangerous icebergs, thereby influencing the weather, and the climate in arctic regions could be considerably improved."

In mid-northern latitudes, according to Oberth, the mirrors could be used to control seasons like spring and autumn to prevent abrupt changes in climate that damage agriculture. He even suggested that entire areas plagued by insects could be decontaminated using the mirrors.

This great German pioneer enjoyed a long life and was therefore able to see many of his ideas become realities. He died at the age of 95 on December 28, 1989, at Feucht, near Nuremberg.

Similar concepts involving clusters of specialized stations were expounded in the late 1920s by two Austrian scientists, Hermann Noordung and Guido von Pirquet. Indeed, mutual influences among Noordung, von Pirquet and Oberth can be seen in their almost contemporaneous works from the 1920s on the same theme.

VON PIRQUET'S "ORBITAL PROJECTS"

Between 1923 and 1929, Baron Guido von Pirquet published articles entitled "*Fahrtrouten*" (Routes for a

Space Voyage) dealing with interplanetary flight and space stations in *Die Rakete* (The Rocket), the journal of the German space-flight association Verein für Raumschiffahrt (VFR). Von Pirquet did not discuss the engineering aspects but detailed the optimum configuration for an orbital base. In other words, he developed what was referred to as "an orbital project" by Willy Ley, a space historian and member of the first group of German pioneers. Von Pirquet took for granted that the stations would be used for a variety of purposes, and he therefore described "a station in three units," each distinguished by its distance from Earth and its function.

The first, "Inner Station" (IS), orbited 760 kilometers up, and the third, "Outer Station" (OS), was even higher—5,000 kilometers above the surface. Both had circular orbits requiring 100 and 200 minutes, respectively, to complete a revolution. The "Transit Station" (TS) moved between the two on an elliptical orbit, forming a link between them rather like a spaceship. In von Pirquet's concept, the station was considered a cosmic support base that could be used as a trampo-

Von Braun's "small station"

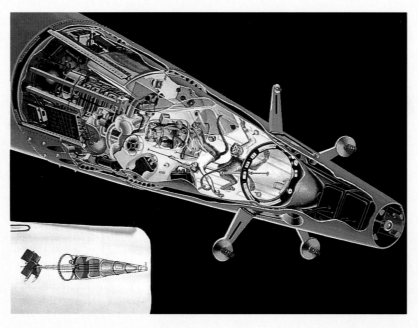

Chesley Bonestell illustration of the "small station" designed by von Braun, orbiting at 320 kilometers, right. The drawing by Fred Freeman at far right shows the interior layout. The plan was to study the reactions of trained monkey astronauts to zero gravity and exposure to cosmic rays.

The station was to stay in orbit for 60 days in preparation for sending human crews into space, and the program to last 10 years at a cost of $4 billion.

(Both illustrations were published in Collier's *in 1953.)*

line, or springboard, for interplanetary voyages and on which the fuel necessary for long trips could be stored.

Baron Guido von Pirquet was born March 30, 1889, in the castle of Hirschstetten, near Vienna. He lived there until 1952, when he moved to a residence in the Austrian capital, where he died in 1966. He studied mechanical, physical and chemical engineering at the Technische Hochschule in Vienna, and in 1926, he was among the founders of the Scientific Society for the Exploration of the Atmosphere.

In 1926, a book by another Austrian, *Das Problem der Befahrung des Weltraums* (The Question of Space Flight), provoked considerable discussion. Although it, too, dealt with rocket propulsion and space navigation, it was in fact the first book almost totally devoted to the different aspects of space stations. The author was Hermann Noordung, who kept his identity secret for decades. The name was actually a pseudonym for Hermann Potocnic, an engineer and former captain in the Austro-Hungarian army.

He was born December 22, 1892, in Pula (now part of Croatia), on the Adriatic Sea. After attending the military-technical academy in Modling, southwest of Vienna, he served in World War I as a railway trans-port technician. After the war, in 1925, he left the army for health reasons and graduated in mechanical engineering from the University of Vienna.

During the 1920s, Potocnic became interested in space and came into contact with Hermann Oberth, to whom he sent his explorations of the subject. Oberth encouraged him to continue and to publish his research. The Austrian engineer followed this advice in early 1929, choosing the pseudonym "Noordung" as a tribute to the group of German pioneers from farther north. A few months later, on August 27, 1929, Potocnic died of tuberculosis without being able to take part in the discussions inspired by his ideas.

His work achieved one result above all others: it brought attention for the first time to a subject that constituted one of the most important objectives of astronautics. Noordung's station design was highly detailed from the point of view of engineering, and the book contained more than 100 drawings with numerous construction details.

NOORDUNG'S THREE STATIONS
Noordung's space station consisted of three components. The first was the *Wohnrad*, or habitation wheel, made up of a series of habitable cells placed in orbit

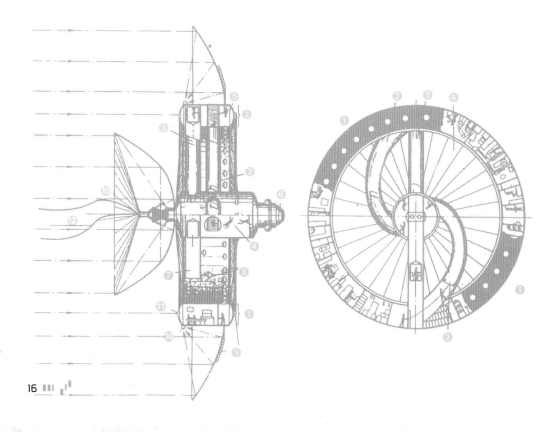

Two illustrations of
Noordung's habitation wheel.

the outer, manned wheel and the central area around the axis of rotation, where there was zero gravity, was by two elevators and two helical corridors.

The first remote station component was the observatory (*Observatorium*). In practice, this was dedicated to scientific activities that could range from research on zero-gravity conditions (the observatory, cylindrical in shape, did not rotate) to astronomical observation. The astronauts reached the observatory after a space walk (perhaps boosted by rocket pistols) and stayed in the laboratory long enough for the required job.

The second remote component was the engine room (*Maschinenhaus*), a large solar concentrator arranged around a module (habitable, if necessary) containing the base's essential electrical and mechanical systems. These included optical and radio telecommunications equipment, batteries to store electricity, and ventilation systems that also served the habitation wheel and the observatory. The latter were linked to the engine room by flexible tubes and cables for energy transport. Nevertheless, emergency services allowed the three components to become temporarily autonomous should the need arise. As a variation in design, the Austrian engineer thought that the engine room could be integrated into the habitation wheel, although in his opinion, this seeming simplification would have complicated the construction.

The uses of the station were traditional and ranged from scientific activities to observation of Earth and the cosmos, from military reconnaissance to iceberg surveillance (against the threat to navigation) and meteorological observation. To achieve these objectives as fully as possible, Noordung placed his station in an equatorial, geostationary orbit at 36,000 kilometers. Consequently, as the station moved in synchronization with Earth, it would appear fixed in the sky at a point that coincided with the Berlin meridian.

When the book came out in 1929, it attracted immediate attention and also some discussion. "Noordung's projects are of great historic interest," wrote Willy Ley to Baron von Pirquet. The latter, however, expressed quite a few criticisms of his fellow countryman. Similar objections were raised in a review in *Die*

separately and then assembled to form the wheel (30 meters proposed diameter), held together by a series of cable spokes attached to a central hub. This was the manned section, with an airlock at one end to allow astronauts to exit and a solar concentrator at the other to collect radiation. The concentrator heated a fluid (liquid nitrogen) which circulated and vaporized to drive turbines connected to an electric motor that generated power. To increase the available energy, Noordung added a ring of concentrators attached to the wheel's surface.

The station rotated on its own axis to create gravity onboard equal to the Earth's. The central airlock and concentrator rotated in a direction opposite to the station's so as to be stable in relation to the surrounding space. This served two purposes. The first was to allow the astronauts easy exit and safe reentry (the dangers of a rotating airlock are easily imaginable). The second was that flexible tubes led from the central concentrator to the two remote components of the station. It should also be noted that access to

Rakete in 1928, a few months before the book's publication. Evidently, the journal had managed to get hold of a copy of the manuscript or a first draft of the text.

The objections raised related to doubts about the efficiency of the main rockets and the location of the station in an orbit so far from Earth. This increased the difficulties and cost of construction, according to the critics, and made its use onerous and impractical. The speed of rotation around the axis (once every eight seconds) to create gravity equal to Earth's was also considered an error as, the critics noted, a lower level of gravity (one-third of Earth's) was enough to be comfortable while at the same time simplifying and lightening the rotation systems.

Also in 1929, a few months after Noordung's book, Hermann Oberth published his second book, into which he had assimilated some of the Austro-Hungarian captain's ideas, of which he was aware thanks to the correspondence between the two scientists. Nonetheless, Hermann Noordung's work is the culmination of the pioneering age of the development of space stations from the point of view of both theory and engineering. It is no coincidence that the concepts put forward after this phase, their intrinsic interest notwithstanding, were either further applications of, or simply references to, Noordung's scheme.

THE 1930S: TOWARD CONCRETE PLANS

The historical events of the 1930s drove researchers in different directions, and the ideas of the pioneers of the cosmos were channeled toward more concrete objectives. To go into space, it was first necessary to build a rocket capable of getting there, and it was on this that all efforts were understandably concentrated.

Experiments in this area were conducted in Russia and in the United States, where Robert Goddard had already flown the first liquid-fuel rocket in 1926. But it was to be Germany that would build the first V-2 ballistic missile, under the guidance of Wernher von Braun and thanks to the interest and relative support of the army.

The son of Baron Magnus von Braun, Wernher was born in 1912 at Wirsitz, when Germany was ruled by Kaiser Wilhelm II. As a student, he became part of the small group of pioneers gathered under the Verein für Raumschiffahrt (Association for Space Flight) that grew up around Hermann Oberth. In 1932, just graduated, von Braun was assigned as technical director of the new research that the army had decided to conduct.

THE PEENEMUNDE SUN GUN

Be that as it may, von Braun's group at Peenemunde on the Baltic Sea saw construction of the V-2 as merely the first step in exploring the universe; and along with building the "weapon of retaliation," they also undertook some studies of future "civilian" developments and the concomitant prospects, including space stations. While at Berlin University in 1931, von Braun himself had written a story entitled *Lunetta*, which described a voyage to a space base.

However, the Peenemunde group also considered the mirror station idea that had previously been put forward by Oberth. Exploring its military applications, they called their version the Sun Gun. It entailed placing a large solar reflector in a 5,100-kilometer orbit, where it would be able to concentrate radiation onto

Arthur C. Clarke's sketch of the three positions for a station in geostationary orbit.

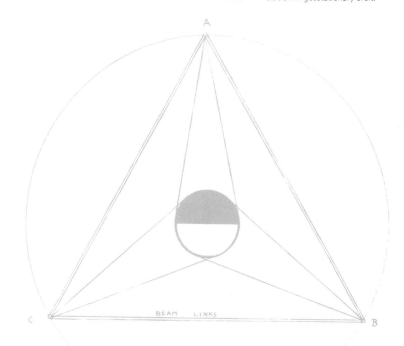

a circular, 64-kilometer-diameter target on the Earth's surface. The idea, a sort of modern version of Archimedes' "burning glasses," was to start fires in enemy cities or induce the boiling of lakes and marine areas. But everything remained only on paper.

CLARKE AND THE TELECOMMUNICATIONS STATION

It was not until the end of World War II that some new investigations of the subject were made. The person to take up the discussion was the British scientist and later great science fiction writer Arthur C. Clarke. In October 1945, he published an article titled "The Space-Station: Its Radio Applications" in the magazine *Wireless World*. Taking up Noordung's concept, he foresaw a new use of the station for terrestrial telecommunications. In more detail, he proposed a system of three bases in geostationary, equatorial orbit 42,000 kilometers from the center of the Earth, each spaced at 120° so that each could cover one-third of the planet. In this way, the antennas would function in relay, allowing almost instant communication with the entire planet. At that time, electronics was not advanced enough to conceive of today's automatic satellites, and therefore, Clarke saw the

orbital station and its astronauts as the right place and instruments for global communication. This concept was later adapted for the geostationary satellites now commonly in use.

ROSS AND SMITH'S BRITISH STATION

Another concept developed from Noordung's project was a space-station proposal published by H.E. Ross and R.A. Smith in the *Journal of the British Interplanetary Society* in 1949. The two British scientists proposed a base whose principal characteristic was a giant parabolic mirror, 60 meters in diameter, to concentrate solar energy to heat a fluid that would activate turbine generators to create the necessary power for the base's functioning. According to their calculations, the mirror would be able to concentrate 3,900 kilowatts of energy that could be converted into a maximum electrical power output of 1,000 kilowatts.

Behind the mirror was a structure of circular areas housing living quarters, research centers and leisure spaces, where 24 people, including two cooks and four servants, could live. British colonies had certain standards to maintain, even in space. Moreover, to provide some comfort, the station rotated on its own axis to generate minimum gravity. A trunnion pro-

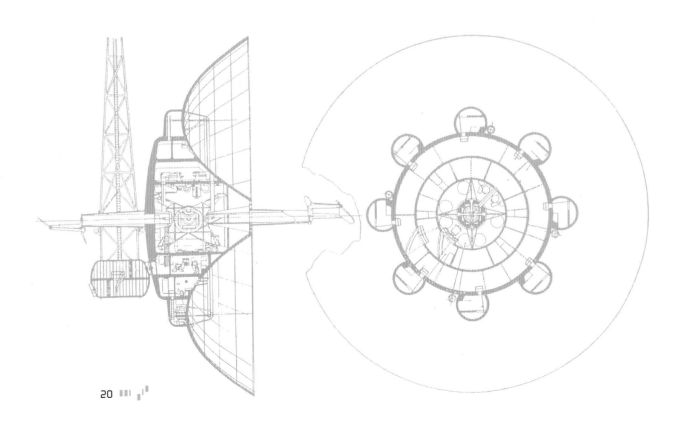

truded from the center of the inhabited zone, with a transverse boom, at one end of which were the radio antennas, and at the other, an isolated laboratory. Assembly was to be done in space after launching the pieces individually. Once completed, the station was to be used for scientific activities and applications and as a platform to build vehicles for interplanetary voyages.

WERNHER VON BRAUN'S WHEEL

In the mid-1950s, the image of the space station mushroomed once more into a dimension halfway between reality and fantasy, between engineering and science fiction. The protagonist in this period, the cul-mination of the pioneering age, was Wernher von Braun, the new icon of the conquest of space.

He had left Germany and had been a resident in the United States since 1945, the end of World War II, working for the U.S. Army first at Fort Bliss, Texas, and then in Huntsville, Alabama. Here, he constructed the new rockets for American defense. At the same time, he planned the next steps to be taken: from the launch of an artificial satellite (which was realized in 1958 with Explorer 1, America's first artificial satel-lite) to voyages to Mars and the construction of a space station.

Shortly afterwards, these ideas were the subject of a series of articles on space travel entitled "Man Will Conquer Space Soon" published in *Collier's* magazine from March 22, 1952. The descriptions were made even more vivid by Chesley Bonestell's illustrations, so much so that the public became greatly interested.

Collier's decision to cover the subject was the result of two conferences. The first—a Space Travel Sympo-sium—was organized in 1951 at the Hayden Planetar-ium by Willy Ley (who had also immigrated to the United States sometime before). Many experts took part—for example, physicist Fred Whipple, who spoke about the upper atmosphere, and Heinz Haber, who dealt with space medicine. Two *Collier's* journalists were also present and were impressed by the speech-es. Some months later, in November, a symposium on space medicine was held in San Antonio, Texas, and

Wernher von Braun was present. Also invited to the meeting was Cornelius Ryan (later to become famous with his book *The Longest Day*), who was, at that time, deputy editor of *Collier's* magazine.

It was on this occasion that Ryan, who at first had been skeptical, became a fervent supporter of the new ideas, thanks to meetings with von Braun and the other scientists. The famous series of articles was born out of this interaction.

One of the pieces, "Crossing the Last Frontier," was devoted to von Braun's space station, which he described as being assembled in a 1,720-kilometer

orbit using a three-stage rocket to deliver the various components. The astronauts then put together the base's prefabricated pieces, inflatable nylon and plastic components, to create a wheel-shaped structure (recalling an old idea of Russia's Tsiolkovsky) 75 meters in diameter with a cylindrical hub in the center. A concave mirror ran round the ring, always pointing toward the Sun to collect and concentrate its energy. By heating and vaporizing the mercury in the central tubes of the mirror, turbines were activated to generate electricity. At that time, it was thought that the absence of gravity would be harmful to the human organism, and so the station was to rotate to create an acceptable level of gravity to accommodate 80 people on the three levels into which the wheel was divided.

According to von Braun, the ideal orbit for the station was a polar one—one that would pass over the Earth's polar regions—because it would make the base "a superb observation point" and, from that orbit, "a trip to the Moon would be merely a second step." The scientist imagined numerous roles for the orbiting base: meteorological surveillance post, point of departure to explore the solar system and a site from which to control peacekeeping. He also saw it as a useful reference tool for air and marine navigation and as a means of transport for atomic bombs. To put these military proposals in the correct light, it should be remembered that when von Braun published them, the United States was still involved in the bitter Korean War.

Finally, the station was not to be an isolated settlement, for von Braun's concept also foresaw a space telescope a short way off that the astronauts could visit, "one or two space vehicles to transport materials" and "space-taxis" linking the transport vehicles with the station. In presenting the articles, *Collier's* pointed out that they were not science fiction stories but real prospects of astronautics.

CLARKE, SCIENCE FICTION AND HOLLYWOOD

The huge interest sparked by the new proposals of the time also touched the imagination of Arthur C. Clarke, this time as a writer. In 1952, he published a science

fiction book, *Island in the Sky*, whose subject was space stations. Two years later, Hollywood got hold of the idea, and Paramount produced the film *Conquest of Space*, whose subject was a space station derived from von Braun's wheel.

Also in 1954, Walt Disney made three films for television to support the construction of new pavilions being built at Disneyland dedicated to outer space and the future. The three films—*Man in Space, Man and the Moon, Mars and Beyond*—were broadcast by the American Broadcasting Company and recycled the content of the *Collier's* articles. They were made with Wernher von Braun, Heinz Haber and Willy Ley as consultants.

The first, *Man in Space*, described the space station proposed by the German scientist. The films began showing in March 1955 and gained a record audience: 100 million viewers. Even President Dwight D. Eisenhower was fascinated by the serious prospects described, so much so that he asked for a copy of the first film that was shown to officials at the Pentagon.

ROMICK, THE LAST PIONEER

While this explosion of half-real, half-fantasy ideas was going on, a new, more technical scheme was taking shape, thanks to a group of engineers at the U.S. Goodyear Aircraft Corporation under the guidance of Darrell C. Romick. After assimilating the concepts expressed in past decades, Romick took a step forward that was considered too innovative for its time. His idea was the construction of a shuttle with three stages, each one manned by astronauts who would pilot a stage back to Earth once its function was ended. In this way, the entire vehicle would be reusable. The next proposal was to use some of the final stages of these shuttles to form a space station by joining them together. The end result would be an imposing, sausage-shaped cylinder about 1,000 meters long with a diameter of 330 meters. At one end, there was a habitation section composed of a rotating disk 15 meters thick and 500 meters in diameter. The whole assembly would accommodate a crew of more than 1,000.

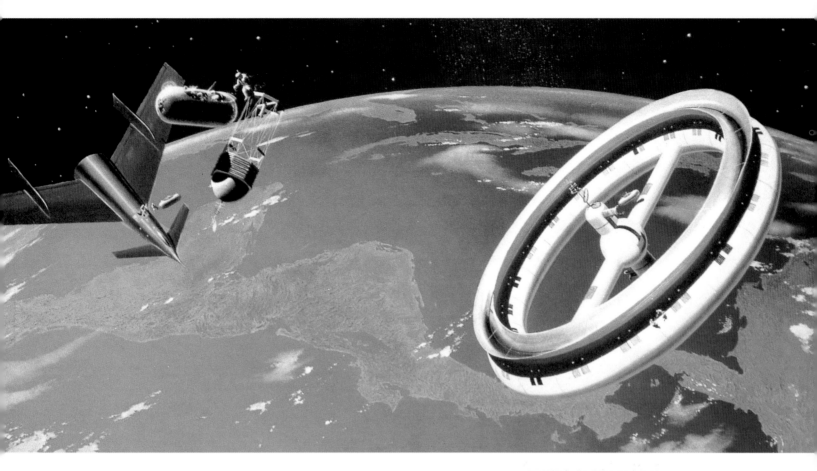

Romick also took into account the economic aspects, which he estimated to be equivalent to the cost of three strategic bombers, while the cost of in-orbit transportation would be $8 per pound (453 grams) of payload. The proposal was therefore quite exhaustive, but despite that, it met with little success. An article published in 1956 in the magazine *Mechanix Illustrated* covered the subject, but its reading remained confined to the closed world of space technicians.

This last imaginative effort capped the long pioneering phase that started at the beginning of the century. By now, in both the United States and the Soviet Union, there was talk of launching the first artificial satellite to mark the International Geophysical Year, an event which duly came to pass on October 4, 1957, when Sputnik 1 inaugurated the space era.

The march of knowledge and necessity also brought the space station into a new and more important phase: actual construction.

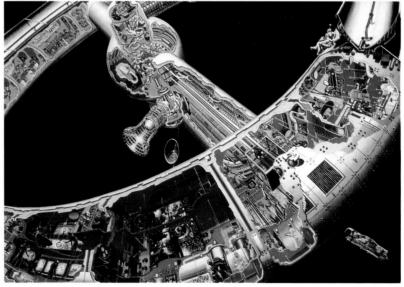

Chesley Bonestell's illustration of Wernher von Braun's wheel station, top. Above, detail of the interior.

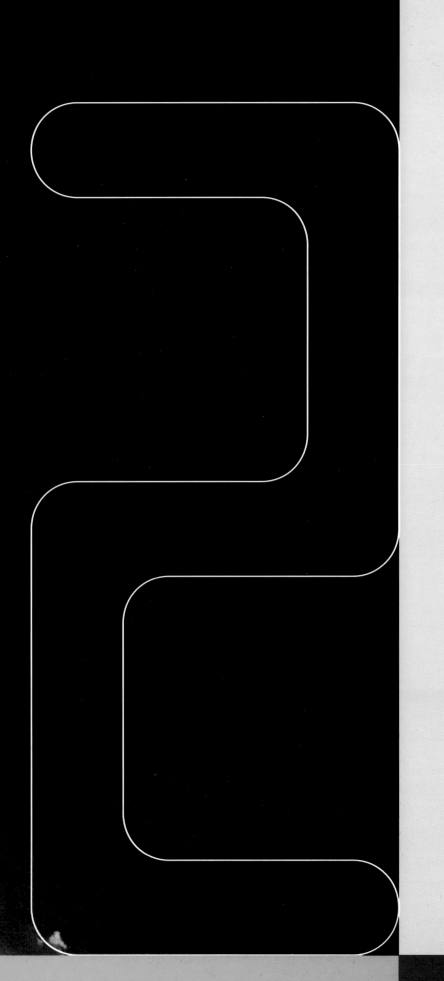

2

The first American station

From civilian and
military projects
to Skylab

The first American station

In the United States, the mid-1950s' enthusiasm for von Braun's ideas, the articles in *Collier's* magazine and the Walt Disney space films were not enough to fuel concrete plans for the construction of a space station. Even the country's simpler plan to mark the International Geophysical Year by sending an artificial satellite around Earth via a small rocket failed to be realized. Moreover, defense circles were not interested in developing long-range intercontinental missiles, despite the new demands of the Cold War. The United States still preferred to use bombers, which could utilize European air bases that were nearer to objectives in the distant Soviet "enemy" territory.

In Moscow, however, the situation was very different. The perceived need to take the threat to far-off America created an opportunity to exploit the new rocket technology whose efficiency (albeit not 100 percent) had been demonstrated with the German V-2 during the final stage of World War II. This desire gave rise to the R-7 vehicle, designed by Sergei Korolev and first successfully launched as an Inter-Continental Ballistic Missile (ICBM) on August 21, 1957. On October 4, 1957, it also carried Sputnik 1 into orbit, the first artificial satellite in history.

That event was traumatic for the Americans, who saw themselves as thereby weakened on the international scene and especially vulnerable in military terms. Their reaction was to make a huge commitment to space in both civilian and defense arenas. Proposals that in previous years had been welcomed as merely fascinating possibilities for the future automatically became projects whose engineering feasibility should be examined.

The space race was on.

On October 1, 1958, the National Aeronautics and Space Administration (NASA) was established, and the following January 1, Congress instituted the House Committee on Science and Astronautics. The committee's first job was to prepare a strategic report, entitled "The Next Ten Years in Space, 1959-69," which outlined the development of the American space effort.

INTEREST IN A MILITARY BASE

Even before the above events took place, however, the U.S. Air Force was looking into the possibility of having a manned base orbiting Earth. In March 1956, following a specific request, it received seven proposals from different companies. The proposed projects were examined, and in early 1958, the Air Force invited the National Advisory Committee for Aeronautics (NACA) to collaborate in their development.

One project design, called the Minimum Manned Satellite (MMS), was considered to have a particularly promising future. Put forward by General Dynamics and Avco Corporation, it suggested using Atlas intercontinental rockets to build a station. The team leader was Krafft Ehricke, a brilliant and visionary scientist who came from the group of German rocket-science émigrés and was now working for General Dynamics. Ehricke's idea was to build the base using three Atlas rockets. Once in orbit, the first rocket's empty tank would become part of the base; the second would

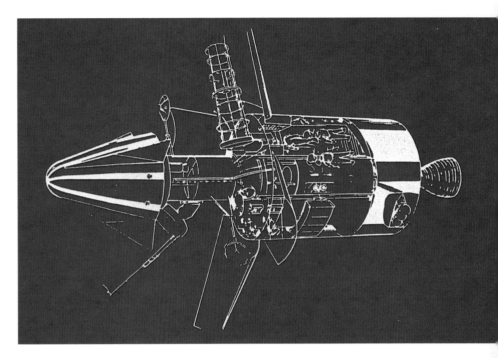

bring supplies and equipment; and the third would have two capsules with four astronauts onboard. Everything would be assembled to form a small space station.

But of all the ideas proposed to take humans into orbit, the U.S. Air Force preferred a simpler plan drawn up by North American and General Electric. However, it was not long before this, too, was canceled; and with the creation of NASA, the Mercury civilian program was chosen to build a capsule able to accommodate only one astronaut.

Plans were also being made to follow Mercury, and at the beginning of 1959, the construction of a space station was considered the proper next step. The same prospect was being examined in military circles, and in less than three months, between March 20 and June 8, General John B. Medaris, chief of the Army Ballistic Missile Agency in Huntsville, Alabama, drew up the Horizon Project. This argued that "military, political and scientific considerations indicate the need for the United States to establish a lunar outpost as soon as possible." It should, he said, be large enough to house 12 people by 1966 and would cost an estimated $6.014 billion. To serve the lunar colony, an intermediate base was proposed—a manned space station orbiting Earth with a crew of 10 astronauts and acting as a supply base for trips to the Moon.

This was the solution that Wernher von Braun had previously explored, and in fact, Project Horizon was born with the participation of the German scientists who were still employed by the Army Ballistic Missile Agency, for which they had built the Jupiter C rocket (the one that carried the first American satellite, Explorer 1, into orbit on January 31, 1958). A few months after the presentation of the Horizon Project, on October 21, 1959, President Eisenhower announced the transfer of von Braun's group to NASA and the transformation of the army's operation at Huntsville into the Marshall Space Flight Center, NASA's new base.

In the first half of 1959, the Research Steering Committee on Manned Space Flight within the newborn aerospace organization was convinced that the

construction of a space station should have priority after Mercury over a manned landing on the Moon. The station and the Moon landing were therefore rival prospects, but according to the committee, both shared the objective of acquiring the necessary know-how for human exploration of the cosmos, which would in future be extended to other planets. In the meantime, the way forward was to build a real space laboratory for experimentation with instruments and crew changes.

These ideas became an integral part of the Long-Range Plan drawn up by NASA's Office of Program Planning and Evaluation and presented on December 16, 1959, which discussed the launching of a manned flight around the Moon and the creation of a permanent space station near Earth in 1965-67. The landing of the first astronauts on the Moon was scheduled for later, in 1970.

LONDON DAILY MAIL'S SPACE-STATION CONTEST
The year 1959 also witnessed an unusual event. The *London Daily Mail*, the British capital's best-selling newspaper, was sponsoring a home show and sought to publicize it as widely as possible. It was decided to announce an international competition among aeronautical companies. The prize was $10,000, and the

The station designed by Douglas Aircraft Company for the *London Daily Mail* competition.

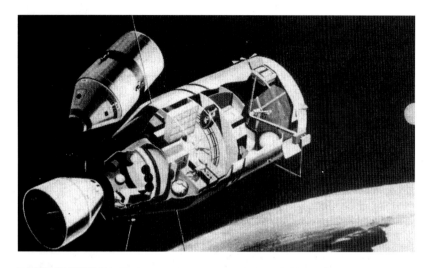

Two representations
of NASA's MORL.

subject was "A Home in Space." The winner was America's Douglas Aircraft Company, which sent a full-scale model of its orbital station-house to London, where it remained on display for two years.

The Douglas "space home" was to be launched into equatorial orbit from Christmas Island in the Pacific. It was a 5-meter-diameter cylinder weighing 13.5 tonnes and terminating in a pointed nose cone. Once in space, the nose would open in three petals that would become solar panels and expose a 60-centimeter-diameter Cassegrain astronomical telescope. At the top was a capsule to be used by a crew of four during launch and return to Earth. The crew would work in the lower, cylindrical section, formed from the vehicle's second stage, now empty after being used to reach orbit. The project generated considerable interest, as the promoters had wished, but it also proved a good opportunity to begin serious examinations of the

notion of a manned orbiting laboratory, and many of the concepts investigated would find applications several years later in the first American station, Skylab.

In April 1960, Los Angeles hosted the first conference devoted to space stations. It attracted experts from many disciplines, and a variety of concepts were presented. One of the most sophisticated was Lockheed's modular station, to be assembled in orbit using a Saturn I vehicle. The station would be powered by a nuclear reactor and have zero-gravity zones for scientific experiments and living spaces equipped with artificial gravity.

But the new prospects were about to receive a setback. On December 20, 1960, during a meeting at the White House, President Eisenhower was presented with a document entitled "Report of the Ad Hoc Panel on Man-in-Space." Prepared by the presidential Science Advisory Committee, the report was to evaluate NASA's plans for the future, which were already in the preliminary budget for 1962. If Eisenhower, always unconvinced about a massive commitment to space despite the frenetic activities in the Soviet Union, hoped for support from his Science Advisory Committee, he was soon disappointed, however, for the report was in perfect harmony with NASA, repeating and underlining the prospect of building an orbital station and preparing a manned mission to the Moon with the Apollo capsules. On reading these unexpected conclusions, the president remained aloof, declaring, according to some witnesses, that he was not disposed to "put his jewels in hock" to send human beings to our satellite, a remark that ironically (and rather unfortunately) called to mind the decision by Queen Isabella of Spain to finance the journey of Christopher Columbus.

KENNEDY IN THE WHITE HOUSE

When the new president, John Fitzgerald Kennedy, moved into the White House, he immediately realized it was imperative to take up the space challenge thrown down by Moscow—a challenge that was becoming increasingly threatening. After putting the first artificial satellite and the first animals into orbit, the Soviets launched a capsule in April 1961 that took Yuri Gagarin into space, the first cosmonaut in history.

NASA's administrator, James Webb, just nominated by Kennedy in January 1961, presented the new president with plans drawn up in the preceding months, suggesting that the best reply to the launch of Gagarin would be to get to the Moon first. Kennedy accepted the proposal, and in May 1961, announced the beginning of the grand Apollo enterprise for the landing of American astronauts on the Earth's Moon by 1970.

The selection of this difficult program and the required concentration of human and financial resources inevitably caused the shelving and postponement of the space-station program—but not its cancellation. In fact, the 1960s became the decade of its deeper exploration, carried out as part of the plans for post-Apollo activities (after human expeditions to the Moon).

Since 1959, one of the NASA bases most involved in the study of space stations had been the Langley Research Center. It summarized the objectives and use of an orbital base according to three aspects: learning to live in space, conducting scientific experiments in the absence of gravity or in artificial gravity; carrying out research into applications for telecommunications and Earth observation; and testing technologies for launching space vehicles from the station.

Starting from these reference points, Langley's engineers examined six possible designs, ranging from a cylinder to a toroid (doughnut shape), all equipped with a rotary motor to generate artificial gravity. The proposed technologies were also investigated, including structures to be assembled in orbit, structures to be deployed once in space and ring-shaped inflatable structures as suggested by von Braun. Comparing the advantages and disadvantages of the various options proved most useful: on the one hand, it limited flights of the imagination, and on the other, it helped lay firmer foundations for the projects selected for development.

MORL, THE LANGLEY CENTER'S LABORATORY

One of the more positive outcomes of these investigations was a study by the Langley center, in collaboration with the Douglas Aircraft Company, which was to become a point of reference in the various stages of the development of a station. Known as the Manned Orbital Research Laboratory (MORL), it was a concept that evolved from three models developed in 1963, 1964 and 1966 dedicated to meeting terrestrial applications, defense needs, support of future space missions and scientific research.

MORL, to be launched with a Saturn I-B, would have a diameter of 6.5 meters. Its three versions could accommodate four, six and nine astronauts, respec-

Seven different space-station models designed by Langley Research Center. The one at bottom left was to be launched in closed configuration, bottom right, and then unfurled in orbit.

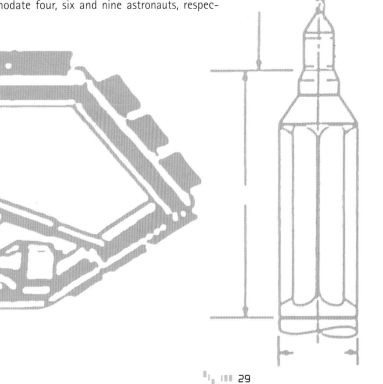

Top, NASA administrator James Fletcher (left) visits President Ford (center) at the White House.
Above, J.F. Kennedy; and the meeting, right, between Brezhnev and Nixon, with astronaut Charles Conrad in the middle.
Facing page, an illustration of the launch of the MOL military station on a Titan 3 rocket: 1) laboratory and capsule assembly, 2) Gemini-B capsule, 3) laboratory. Inset, the capsule with the two astronauts onboard separates from the laboratory for return to Earth.

to be attached by cables to the final stage of a Saturn rocket. By rotating the two together, a level of gravity equal to one-third of the Earth's could be created. This and other proposals were collected in a series of documents entitled "Blue Books," which were later to prove invaluable.

HOUSTON'S OLYMPUS PROJECT

Like the Langley center, the Manned Spacecraft Center in Houston, Texas, and the Marshall Space Flight Center in Huntsville, Alabama, were also working on space stations. In this context, Houston produced the Olympus Project. Very similar to MORL, it was to be a cylindrical station also made up of compartments and offering zones of reduced gravity and zero gravity. Taken into orbit by a Saturn V, it would be large enough to accommodate 24 people. One of the different possibilities considered was its use as an orbiting base for interplanetary missions.

At Huntsville, the center directed by Wernher von Braun, more pragmatic research was preferred and was to prove successful. Their idea was to create a space station inside the Saturn IV-B (the upper stage of the Saturn V used for the lunar missions), exploiting its empty tank space.

All these initiatives by the NASA bases did not, however, induce the military to abandon its idea of having its own cosmic observation post. The U.S. Air Force had already proposed the Blue Gemini Plan to use the capsules for defense purposes at the end of NASA's Gemini program. But in 1963, Congress turned down this proposal—even after the suggestion was made of adding a small laboratory to broaden the restricted possibilities of the capsule—arguing that the high costs were not offset by the expected advantages. After only a few months, an improved idea was proposed by the same defense secretary, Robert McNamara, during the December 10 meeting of the National Aeronautics and Space Council.

MOL, THE PENTAGON'S STATION

During that meeting, the plans for the military mini-vehicle Dyna-Soar, which had evolved over recent

tively. The first two versions were to be launched in 1968 and 1970 and placed in a 360-kilometer circular orbit inclined 28° with respect to the equator. The third and largest version, whose launch was planned for 1972, would have a lower orbit with a higher inclination (50°) to cover a wider area of the planet. MORL would have zero gravity, but there would be a centrifuge to simulate return to Earth and to control the physical conditions onboard. With a "shirtsleeve" environment onboard, the station would have compartments, one above the other, each with a different function. At the top were the astronauts' quarters; below them, the centrifuge; then the control center; and finally, the scientific laboratory. In addition, there was a storage compartment for materials brought from Earth. Power was to be supplied by solar-cell panels, but a Brayton isotope cycle system was also considered.

In one of the last versions to be studied, MORL was

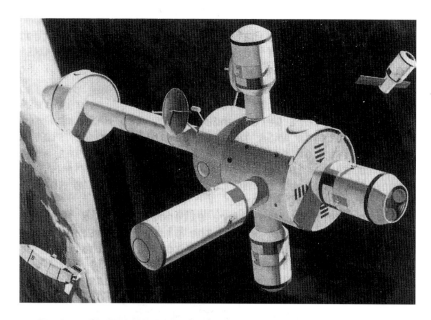

NASA station designs from 1978 for 12 people, top, and from 1983 for 50 people, above.

the defense secretary continued, "a manned orbiting laboratory, not for a precise, clearly defined and identified military mission, but because we believe it useful to develop some technologies that could prove essential for manned military operations in space."

The following month, January 1964, the U.S. Air Force assigned some contracts for studies christened OSSS (Orbital Space Station Studies) to define the characteristics of the future laboratory, specifying configuration, dimensions, life-support systems, crew composition, mission duration, facilities necessary for the development of the program, test equipment and recovery techniques. The studies were the continuation of two previous research projects: the Military Test Space Station (MTSS), conducted from 1958 to 1961, and the Military Orbital Development System (MODS), begun in 1962.

On August 25, 1965, President Lyndon Johnson, who came to office after President Kennedy was assassinated, gave his official approval for the MOL project. The Douglas Aircraft Company, which had previously worked on von Braun's proposal at the Marshall Space Flight Center for a station to be built from the empty final stage of a Saturn V rocket, was chosen to design and build the laboratory.

The birth of this military project and the real beginning of its construction—even though it would remain unfinished—was significant for the almost immediate consequences it provoked outside the United States, in the Soviet Union. The MOL was to be launched in 1969 and would permit two astronauts to remain in orbit for at least 30 days in shirtsleeve conditions. It was composed of a Gemini capsule (called Gemini-B) minus the original service module used when it flew for NASA. This would leave open a hatch in the heat shield at the base of the capsule to allow astronauts to enter the pressurized laboratory attached thereto. The laboratory was 3.05 meters in diameter, 12.5 meters long and weighed 9 tonnes. The overall capsule-laboratory assembly was 15.2 meters long and weighed 11.8 tonnes.

A new version of the Titan III vehicle, called III-M, was needed to take it into space. In practice, this was a

years, were canceled, and the MOL (Manned Orbital Laboratory) program was given the go-ahead. The project was known as the "Gemini-X plus laboratory," because it was to take NASA's Gemini two-seater capsule and modify it to allow astronauts to pass from the capsule to the laboratory. "The objective of Gemini-X," McNamara emphasized, "will be to test operations in space using equipment and personnel that can also meet some military needs. We propose,"

Titan III-C with its two lateral solid-propellant boosters lengthened (seven elements instead of the usual five). This served to increase its capacity, enabling it to lift the MOL load into space.

The ideal launch base was Vandenberg, in California, from which it was possible to reach the necessary polar orbit to cover the entire planet. (For security reasons, during the first phase of flight, it was not possible to go over 57° latitude with a launch from Cape Canaveral, consequently limiting reconnaissance.) All the same, to give better guarantees during the launch from Vandenberg, the Department of Defense bought a piece of land called Sudden Ranch in the area south of the base.

The initial plan was to build seven MOL laboratories, the first two to be launched unmanned to check out their systems. Launches would take place every six months. Subsequently, the manned missions were reduced to four (one being canceled for economic reasons). Although the cost of the program was originally estimated at $1.63 billion, it had risen over the years to $3 billion, and the date of the first launch had been postponed to 1972.

While the design was taking shape, 17 astronauts were selected to fly in the laboratory. Of these, 13 were from the Air Force, 3 from the Navy and 1 from the Marine Corps.

And so on November 3, 1966, a Titan III-C launched the first suborbital flight of a mock-up of the MOL to test the stability of such a long payload and to check out the performance of the modified Gemini capsule during reentry. The test was successful, and the program proceeded along with advances in reconnaissance and data-transmission techniques that would allow the making of real-time decisions during space flight.

But two circumstances, one technical and the other political, were leading to a decision against the MOL. The first concerned the development of technologies that permitted spy satellites to be built, replacing manned reconnaissance vehicles as well as being less expensive precisely because they did not have to safeguard the lives of astronauts. The second circumstance was linked to the escalation of the war in Vietnam, which was absorbing more and more of the defense budget, forcing cuts in other, not immediately necessary, spending.

North American Rockwell Space Division's concept for a 12-person station drawn up in 1969 following NASA's selection of two companies for the design project.

On June 10, 1969, while NASA was preparing for an imminent landing on the Moon, the Department of Defense announced the cancellation of the MOL program, for which $1,495 billion had already been spent. While it is true that this decision was reached for the reasons given, according to some experts, the primary negative aspect of the program was a lack of clarity in defining the mission that rendered it subject to delays and inevitable cost increases.

THE OBJECTIVES FOR A CIVILIAN BASE

While the military was discarding the idea of using humans in space in favor of robotic systems, NASA was demonstrating increasing interest in building a permanently manned orbital base. This interest was integral to the Apollo Extension System (AES) plan, later renamed Apollo Application Program (AAP), begun in late 1964 as a logical follow-up to the lunar missions. The studies conducted in previous years were therefore continued, and a further step forward was taken. For the first time, the projected station was dealt with in an organized and in-depth manner with the undertaking of a two-part project, one component of which was to design the entire system, the

A 1969 station design by NASA.

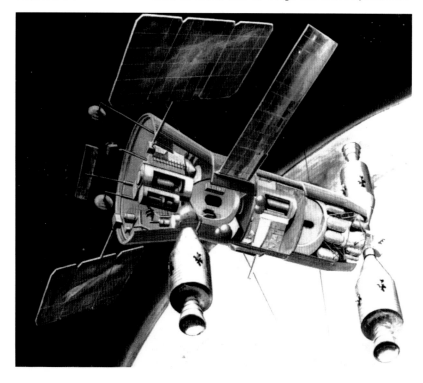

other to define its characteristics in order to meet eight objectives considered necessary to justify the construction of the station itself.

Those objectives were as follows: astronomical research with optical and radio telescopes, geographical observation of terrestrial resources, development of meteorological sensors, biomedical research, aeromedical research, development of advanced technologies, development of flight operations to render the missions more efficient and, finally, making long-duration stays to prepare for future human missions to the planets.

Both the design component, led by E.Z. Gray, and the characteristics-definition component, under Charles Donlan, came to the conclusion that the space station represented an important opportunity to learn to live in and utilize the possibilities offered by the space environment. It would, therefore, be an essential tool for scientific and technological research. The study compared and contrasted the design of the "single environment" station with one having different components dedicated to various activities. "The most difficult problem," Donlan emphasized, "is artificial gravity. If it becomes an indispensable requirement for the astronauts' comfort, its generation will have a great impact on the design of the base, as the majority of the experiments to be conducted onboard require the absence of gravity." Other conflicts to solve concerned the desire to conduct celestial and terrestrial observations at the same time.

In conclusion, the project maintained that "a *minimum station*" would have to meet many of the identified needs. It would operate in a 360-kilometer orbit inclined 50° to the equator. It would also have to operate full-time or for sustained periods over a five-year span and be large enough to accommodate a staff of 8 to 12.

NASA CHOOSES A STATION MODEL

With the support of the Donlan-Gray study, NASA decided to proceed with the station program, requesting for the first time specific funding in the budget for the 1967 fiscal year. Unfortunately, the Bureau of the Budget did not approve this request; but NASA went

Astronaut-geologist Harrison
M. Schmitt on the Moon
during Apollo 17, the last
manned lunar mission.

ahead anyway, using internal funds (albeit reduced) for 1967 and 1968.

The aim was to reach 1969 with sufficient evaluation completed to allow continuation of the program through Phase B, dedicated to detailed studies. In this way, the space administration, Congress and the White House would have all the information necessary to give a definitive go-ahead to the station.

In the year of the lunar landing, NASA was still unsure of how to proceed, despite having carried out much research. What was certain was the intention to test the endurance of humans and systems for long periods in orbit, to find out whether the astronauts would be capable of the necessary scientific and tech- nological activity and to understand how to ensure continuity of life on the base through crew changes and resupply missions. Meanwhile, costs, construction options and security levels were analyzed.

On April 19, 1969, NASA finally gave the go-ahead to Phase B, laying down the guidelines that industry should follow in presenting its proposals. The objec- tive was a station constituting a "general-purpose laboratory in Earth orbit for scientific and technolog- ical experiments to obtain useful applications and further developments of the capability to explore space."

The base was to be a 16-meter-long cylinder with a diameter of 10 meters, built to accommodate 12

Right, cutaway view of Skylab showing its two habitable levels: upper for working and lower for living. The large container at the base was for refuse.

astronauts. Supplies would arrive from Earth every 15 days, and the crew would be changed every 90 days. The selected orbit was at an altitude of 480 kilometers, either inclined 50° to the equator or passing over the poles. The internal environment was to be zero gravity, but a mass was to be attached to the base by a cable, and by moving this away and rotating the two bodies, artificial gravity could be generated in the station to study the effects. With admirable foresight, the guidelines also added that the model to be designed should not be just the fundamental element of a future large orbital base but also the core of a future vehicle for the exploration of Mars.

FIRST SPACE CONTRACTS

NASA offered two contracts for the study, each worth $2.9 million. Three corporations responded: North American Rockwell, McDonnell Douglas and Grumman Aircraft. On July 22, 1969, North American and McDonnell were chosen to work, respectively, under the direction of the Johnson Space Center in Houston, Texas, and the Marshall Space Flight Center in Huntsville, Alabama.

Meanwhile at NASA headquarters in Washington, groups and offices (Office of Manned Space Flight, Space Station Steering Group and Space Station Task Force) were formed to supervise the new field of activity, overseen by an independent organization called the Space Station Review Group. Pursuing a policy of

openness toward friendly countries, the space administration invited representatives of foreign industry to the periodic meetings dedicated to the program, in the hope that other nations would join the undertaking. In the meantime, a launch date of 1977 was projected, together with an estimated overall cost of $8 billion to $15 billion, including 10 years of in-orbit operations but excluding the costs of transportation into space.

As the corporations developed the details, the station design became more and more precise. Most of its internal volume was divided into four floors: two for the laboratories and two for the crew. A fifth floor was reserved for the base's energy system and other equipment. It was also thought that external modules could be added, if necessary, at a later date. The two solutions studied by the industrial competitors differed mostly in the systems adopted (for example, using solar or radioisotope generators as a power source). Likewise, plans were made to stimulate interest among possible future international users of the base.

FROM SATURN V TO THE SHUTTLE: THE STATION CHANGES DESIGN

On July 29, 1970, NASA management ordered its Houston and Huntsville bases to terminate Phase B. At the same time, it also ordered termination of the Saturn V rocket program, used up till then for the lunar missions. This caused a crisis in the plans for the station, for the studies conducted thus far were linked to the large launch vehicle, which allowed a much greater diameter of 10 meters for the station.

An inevitable sense of frustration descended on the NASA bases and the companies who had been working on design solutions that were now almost unusable. The order, from that moment, was to use the space shuttle that was then in development and whose cargo bay could hold the pieces of the station to be launched. Now, the maximum permissible dimensions were equal to a cylinder 18 meters long and 4 meters wide—that is, less than half of what had been expected.

Despite the change, NASA still considered it imperative not to alter the capacity to house 12 crew members, with some space made available for the laboratories. This meant a general review, and the new terms were contained in a document published by the Manned Spacecraft Center in Houston in November 1970. This no longer talked of a basic module with certain possibilities but of a modular station, in pieces to be assembled with several shuttle flights. The document stated that "the modular station will consist of modules to be launched separately and then assembled in an orbit between 445 and 500 kilometers with an inclination of 55° to the equator. The Initial Space Station will be large enough to accommodate a crew of six astronauts in a first module to be launched in January 1978. Subsequently, it may grow to a configuration sufficient to accommodate 12 people as originally defined. The new approach of a type of station that will grow in time should lead to a reduction in the initial funds necessary, while the station should reach full capacity in 1984."

North American Rockwell and McDonnell Douglas reviewed their plans, and the new Initial Space Station that resulted was formed of three habitable modules (one for the crew, one for the general laboratory and a logistics module, a sort of storehouse for instruments and various equipment.

Another module was added for power generation using solar-cell panels. The new concept of modularity stimulated the imaginations of the designers in these months, and they produced a highly evolved plan for a station with almost 20 modules.

PRESIDENT NIXON REQUESTS A REPORT

But reality unfortunately did not live up to expectations. In early 1969, President Richard Nixon moved into the White House, and he formed a Space Task Group to redefine what had until then been called the Post-Apollo Program, a follow-up to the enterprise that in July of that year had put the first Americans on the Moon.

Skylab's living quarters:

1) eating area

2) toilet

3) sleeping area with three cabins, one for each astronaut

4) area for physical activity and medical checkups

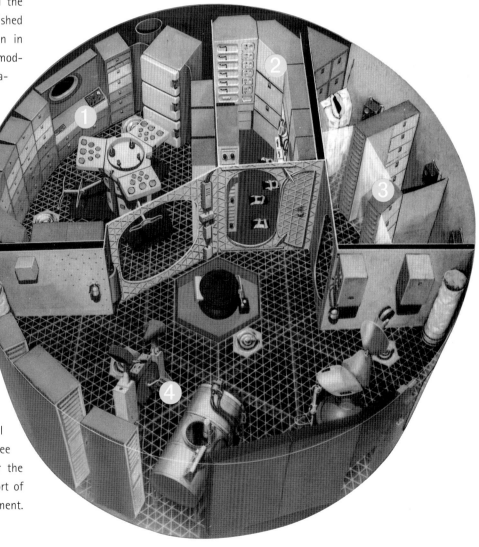

Schematic drawing of Skylab
with its various components:
1) ATM (Apollo Telescope
Mount) solar observatory
2) solar panels
3) antimeteorite shield
4) experiments
5) sleeping cabins
6) eating area
7) storage
8) laboratory area
9) Airlock Module
10) reserve docking system
11) main docking module
12) Apollo capsule
Skylab was 35 meters long
and weighed 76 tonnes
(without the Apollo capsule).
Its widest diameter was 6.7
meters.

The group was presided over by Vice President Spiro T. Agnew and was made up of NASA administrator Thomas O. Paine, Air Force secretary Robert C. Seamans, the president's scientific counselor Dr. Lee A. Dubridge and three "observers"—Under-Secretary of State U. Alexis Johnson; Glenn T. Seaborg, president of the Atomic Energy Commission; and Robert P. Mayo, director of the Budget Office. The group presented its report, "The Post-Apollo Space Program: Directions for the Future," in September 1969. Although it supported the necessity to continue human exploration of the Moon, it also made a priority of starting two large programs: a space station above Earth and a reusable shuttle. The next steps would be a space station orbiting the Moon, a lunar colony and, finally, a human landing on Mars "before the end of the century."

VON BRAUN AT NASA IN WASHINGTON
Meanwhile, NASA administrator Thomas Paine called Wernher von Braun, the great inspirer of America's space program, to Washington. So, in February 1970,

von Braun left the direction of the Marshall center in Huntsville to become NASA's vice administrator for future planning.

While this was going on, however, the American public's enthusiasm for space exploration was waning after the first lunar landings, and there was growing criticism of space-program activities, accompanied by the demand for greater attention to terrestrial problems. As a politician, President Nixon demonstrated his sensitivity to these new views. Moreover, it was now clear that the political-military objective of reestablishing American superiority after the Soviet Union launched the first artificial satellite had been fully achieved with the conquest of the Moon. The inevitable consequence was a reduction in the space budget and a downsizing of its programs.

This was already felt at NASA in 1970, and the efforts to maintain a strong post-Apollo program as recommended by the Space Task Group appeared increasingly futile. Even Wernher von Braun was of the opinion that if a choice had to be made between building a reusable space shuttle and an orbiting space station, the first was certainly more useful, because without it, the proper use of the station was unfeasible.

Nixon chose the shuttle and abandoned the large station. Thus, although plans for an orbital base continued into 1971, in January 1972, when the president formally initiated the space-shuttle program, all other plans were automatically shelved, there being insufficient resources for them to proceed. All that remained of the idea for a station was the proposal to build a laboratory module to fit inside the cargo bay of the shuttle, on which it would depend completely, allowing scientific missions to be run for a few days and with the presence of scientist-astronauts.

NASA therefore abolished the Space Station Task Force and replaced it with the Sortie Lab Task Force. A result of this new direction, which also foresaw Europe's involvement in the shuttle program, was the birth of the Spacelab laboratory. This would be built by the European Space Agency (ESA) and then handed over to NASA to be used in those shuttle missions for which it was considered necessary.

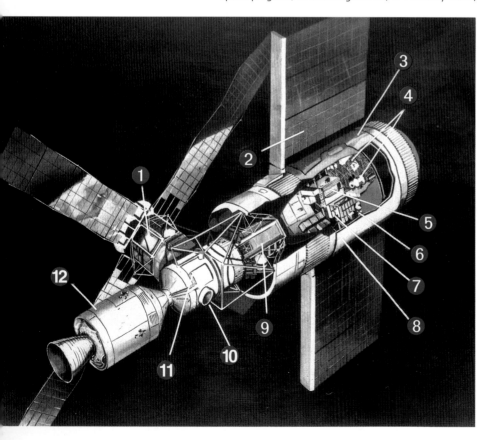

THE BIRTH OF SKYLAB,
THE FIRST AMERICAN STATION

The decision to build the reusable shuttle had halted the plans for a large manned orbital base. As it happened, though, another project was fully under way at NASA that was to give the United States its first space station, albeit with more limited objectives. This meant that all the research conducted with a more ambitious end in mind had the chance to be applied and confirmed.

By late 1964, a broader use was already being sought for the enormous amount of technology, systems, instruments and apparatus developed during the Apollo lunar program. To aid in this objective, on August 6, 1965, NASA opened the Apollo Application Office at its headquarters in Washington. The office began to work on an idea suggested in the early 1960s of installing a solar observatory (Apollo Telescope Mount) in the service module of the Apollo spacecraft. Once it had been used to make observations, and after the data and images had been recovered, it would be abandoned together with the service module that had been separated from the Apollo capsule for the return to Earth.

In March 1966, some ideas started to take shape about using the Saturn IV-B—until then used as the second stage of the Saturn I-B rocket and the third stage of the lunar rocket Saturn V—as a laboratory. Once the hydrogen and oxygen fuel in the rocket's

The Cape Canaveral mission control center at the launch of Skylab.

Stations in science fiction films: *2001, A Space Odyssey*

Stanley Kubrick's 2001, A Space Odyssey was released in 1968. Based on the book of the same title by British writer and scientist Arthur C. Clarke, the film is about a voyage through time and space to try to make sense of evolution, as symbolized by a black monolith found on the Moon. A signal emitted by the monolith points toward Jupiter, and a Jupiter mission sets out to explain the mystery of the monolith.

A Saturn V rocket lifts off
from Cape Canaveral carrying
Skylab on May 14, 1973.

tanks had been exhausted, an Apollo capsule could dock with the empty stage to allow the astronauts to enter. In this case, it would be used only as an empty vehicle in which the astronauts could function with some protection while conducting observations.

A second solution was to transform the empty stage into an orbiting laboratory with equipment brought from Earth. In this case, the stage would be launched with an additional module (Multiple Docking Adapter) equipped with two possible hatches for docking. The Apollo capsule would be attached to one of these. Finally, there was also an idea for the separate launch of a solar telescope joined to the upper part of an LEM (Lunar Excursion Module, used in the lunar landings). Together, they would then dock with the second, free hatch, and the astronaut-scientists would have to move into the LEM cockpit to work with the telescope.

In any case, the laboratory—except for the telescope, which had its own power source—would have to function with energy supplied by the Apollo command module. This was a significant limitation and prompted the addition of solar-cell panels to increase power available for the mission. Thus, while a "cluster" concept of different modules to be joined together was developing, the debate between those who supported the idea of creating a laboratory in orbit from an empty rocket stage (the "wet solution" because it would start with a rocket tank filled with fuel) and those who were convinced that it would be better to launch a fully equipped laboratory inside the Saturn IV-B stage (the "dry solution," without fuel) was becoming increasingly lively. The NASA bases directly involved were the Johnson Space Center in Houston and the Marshall Space Flight Center, directed by von Braun, in Huntsville.

JULY 1969: GO-AHEAD FOR SKYLAB

On July 22, 1969, in the wake of the successful landing on the Moon two days earlier, NASA announced that an orbital laboratory linked with a solar telescope would be launched on a Saturn V rocket. The Saturn IV-B stage would be customized before launch to make it into a laboratory with instruments and

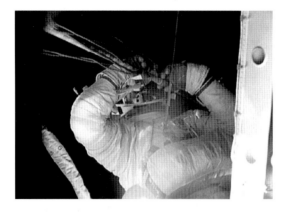

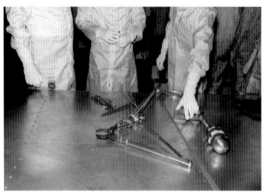

systems to permit the astronauts to live onboard: the "dry solution" had prevailed. The announcement mentioned two orbiting laboratories, the first of which would leave in July 1972, plus seven Saturn I-B launches for the astronauts. The plan was finally given a name on February 24, 1970, as the Apollo Application Program became Skylab.

But in the same month, budget restrictions imposed by the circumstances of the times, especially the war in Vietnam, led NASA to announce cancellation of the last planned mission to the Moon, Apollo 20. The following September, another two, Apollo 18 and 19, were cut. Skylab could not avoid the same problems, and in 1971, while development and construction were beginning, it was decided to build only one unit; its launch was postponed by nearly a year to April 1973, and there would be only three crew launches, each with three astronauts.

Responsibilities were shared between three NASA centers: Marshall in Huntsville, under von Braun, to oversee the construction of the components (laboratory, airlock, multiple docking module, solar observatory, onboard instruments) as well as having responsibility for the Saturn launchers; Johnson in Houston for mission management (including recovery of the data and specimens and their analysis), training of the astronauts and development of the required modifications to the Apollo command and service modules; Kennedy for mission control and launch of the laboratory and crews, with a clause requiring the ability to make two launches in 24 hours, if necessary.

Naturally, supporting the three NASA centers, numerous other administration establishments, universities and corporations were involved, all well integrated to ensure, first, the completion of the program and then the essential support during missions. This complex network of men and means would soon have the chance to prove its efficiency; and fortunately, it did so.

THE OBJECTIVES OF THE
FIRST ORBITAL LABORATORY
The construction and use of Skylab had four objectives. The most important was to investigate human

performance and endurance during prolonged space flight by analyzing basic biomedical processes. Until then, the longest stay in orbit had been a couple of weeks with the Gemini two-seater capsule, in which the astronauts could not even move. The future held lunar voyages, orbiting stations and a landing on Mars. It was, therefore, necessary to understand the human organism's reactions and behavior in the hostile space environment, particularly in the absence of gravity, the fundamental element governing human physiology. How would this affect, for good or ill, blood circulation, metabolic processes, cell mechanisms, tissue growth and sensory functions? Moreover, the psychological aspects needed to be studied, as well as the potential and limits of the astronauts, who had to be explorers, technicians, scientists, pilots and navigators, even doctors, all at the same time.

How automatic should the orbital vehicle or laboratory be? How much should be left to the direct management of its inhabitants? These were the first questions to be asked and the first pieces of knowledge to be gathered to build future vehicles and space settlements, ensuring humans safe and comfortable survival and the capacity to work productively.

Second, Skylab had to permit the development of efficient sensors and techniques for environmental and resource observation of Earth that could be done best from space. It must also permit astronomical observations devoted to the Sun, whose continuous energy output directly influences life on Earth, from the atmosphere to the human body. Last, it must enable research into the creation of new materials (metal alloys, semiconductor crystals, pharmaceuticals) permitted by the absence of gravity and impossible to obtain on Earth.

THE SKYLAB CONCEPT

Skylab allowed the application of a wide range of technologies developed for the Apollo lunar program, which thus yielded further, fruitful use. These included things from propulsion to navigation and guidance systems, from structural engineering to environmental control devices, from sensors to data-handling

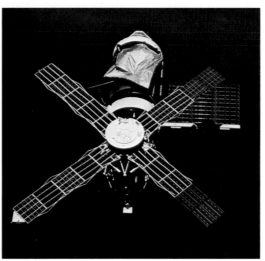

Two photographs of Skylab taken at the end of the first crew's stay: the solar panel is deployed, and the parasol is open.

systems and from power generation to telecommunications. Moreover, the design of Skylab permitted the use of already developed vehicles like the Saturn I-B and Saturn V and the Apollo spacecraft.

But this was merely the starting point for the program. In fact, the engineers had to adapt all the available systems to render them efficient when used over a long time span and to meet the diverse requirements of a space settlement that must be operative and safe for at least nine months. In the Apollo program, these systems had had a more limited use of around two weeks. The difference was therefore considerable, raising previously unknown problems and requiring new safeguards. Naturally, the different requirements, in any case, demanded the development of some new technological equipment.

The program also yielded a chance to apply the previously mentioned vast range of research carried out in the 1960s toward an orbital base, wherein the policy drawn up by the Marshall center in Huntsville, directed by von Braun, had prevailed. The result was a station with four modules: the Orbital Workshop, the Airlock module to allow the astronauts to egress, the Multiple Docking Adapter and the Apollo Telescope Mount. In addition, there were the Apollo command and service modules, with which crews would arrive at the station and return to Earth at the end of the mission. As stipulated, a Saturn V would be used to take Skylab into orbit, and the astronauts would leave with the smaller Saturn I-B.

THE STATION MODULES:
THE ORBITAL WORKSHOP

The heart of Skylab was the Orbital Workshop laboratory, which was created by transforming a Saturn IV-B (usually used either as the third stage of a Saturn V rocket or the second stage of a Saturn I-B). The 6.7-meter-diameter structure remained unchanged, as the volume occupied by the large liquid-hydrogen tank became the work area (the laboratory), with a lower section separated by a floor grating and given over to the astronauts' living quarters. Lower still was the small liquid-oxygen tank, now used as a container for refuse accumulated during the missions. The habitable volume, including laboratory and living area, was 275 cubic meters, approximately equivalent to a 90-square-meter apartment.

To protect the capsule from meteorites, the whole cylindrical body of the station was covered by a shield of thin aluminum sheeting separated by 15 centimeters from the main exterior wall, which was also aluminum alloy, made of hollow panels 1.9 centimeters thick. If, as was statistically possible, small meteorite particles hit the shield and penetrated it, they would thereby break up into smaller ones that would not damage the laboratory wall. During launch, the antimeteorite shield lay right against the laboratory structure, and it had to be lifted into position by a torsion bar once in orbit.

Living area. The inside of the astronauts' living area was divided by walls to separate the eating zone, with its fixed table, from the sleeping area, with individual cubicles for each astronaut to have a little privacy, and to divide the bathroom area from the zone for medical experiments. The bathroom equipment was fixed on one wall, with a suction system to replace gravity for collecting feces and urine. Samples were taken of both (by drying the feces and freezing the urine) to be brought back to Earth for analysis of the effects of space conditions, diet and workload. The remainder was put in a sack in the waste tank. When an astronaut sat on the space toilet, a safety belt held him in place.

The area for medical experiments had three bulky instruments: a chamber to expose the lower half of the body to negative pressure (to move more blood down to the legs as gravity would do on Earth) and to test how the cardiovascular system adapted to the absence of gravity, an ergometric system (exercise cycle) and a rotating chair for tests of the vestibular apparatus. A mobile shower of waterproof material

Below, the first Skylab crew (from left to right): Joseph P. Kerwin, Charles Conrad and Paul Weitz.
Above, Skylab mission patch. On facing page, astronaut Alan Bean collecting films from the solar observatory.

distributed a fixed amount of water and soap, which was then recovered by a suction system. In the middle of the ceiling was a circular hole leading to the upper area, dedicated to technological experiments.

Work area. This zone also housed water reserves, food supplies, three refrigerators and an environmental control system that distributed air consisting of 74

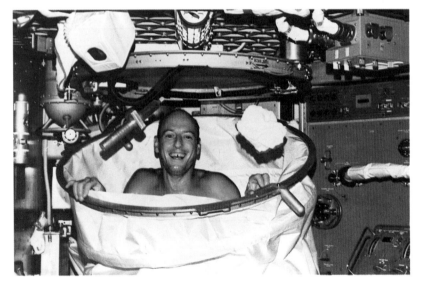

percent oxygen and 26 percent nitrogen through a network of conduits.

The environment throughout Skylab had a lower atmospheric pressure than Earth (about one-third) and allowed the astronauts to live in shirtsleeves. The temperature could be regulated between 13° and 32°C, with a relative humidity of 26 percent at 29°C. The atmosphere was continually purified: carbon dioxide was removed by two zeolite plants, and bad smells were absorbed by carbon filters. In addition, a

layer of insulation (polyurethane foam) on the internal surfaces protected the environment from changes in temperature when the laboratory was exposed to the Sun or when it flew in the darkness of night. The insulation also served as protection from intense ultraviolet radiation. Moreover, all the materials used were extremely stable and could not emit volatile particles or gases.

As the atmosphere was rich in oxygen and therefore a fire hazard, the majority of the materials, including clothes, were fireproof, and those which could catch fire were clearly marked. The laboratory walls had two airlock windows (one facing the Sun and one on the other, shaded side) through which it was possible to expose some scientific instruments to the vacuum of space. A variety of instruments and equipment was stored onboard for different needs or emergencies. Among these were repair kits, film, cameras, recorders, pincers, pliers, scissors, clasps, thermometers and emergency lights.

Handrails ran along the walls to help the astronauts move about, and at certain points, especially near the instruments, foot restraints helped an astronaut stay in position. Finally, radiators were placed along the walls and activated according to the requirements determined by the laboratory's position.

The Saturn V guidance unit. The upper part of the laboratory ended in a ring whose inner wall housed all the guidance and control systems of the Saturn V rocket, controlling it from lift-off from the ramp at Cape Canaveral until orbital insertion of the third stage, which was in fact the Saturn IV-B, transformed in this case into Skylab.

Once in space, the "brains" of the big rocket also controlled all the operations of the station (pressurization, activation of the refrigeration systems, deployment of the antimeteorite shield, and so on), including placing it in the desired position. The same system would open the solar panels of both the laboratory and the telescope, although those operations could also be commanded by the controllers in Houston. After 6.5 hours, the batteries of the Saturn V's "brain" would be run down, and its function would cease completely. At that point, Skylab would manage itself with its own onboard systems.

THE AIRLOCK MODULE

A cylindrical module, the airlock, 5.4 meters long and 3.1 meters wide, was attached to the top of the laboratory to provide access to space. The station's control center for electrical, environmental and telecommunications systems and the automatic alarm systems were also housed here. Its role was, therefore, extremely important. The cylindrical oxygen tanks and spherical nitrogen ones to feed the internal atmosphere were arranged on its exterior.

At either end of the airlock was a door (the same

type as used in the Gemini capsules), which was closed when the astronauts, wearing space suits, were about to go out for a space walk. The airlock was then depressurized, releasing the outside door.

THE MULTIPLE DOCKING ADAPTER

While one side of the airlock contained the laboratory, the other housed the multiple docking adapter, a cylinder 5.1 meters long with a diameter of 3 meters. This module was actually a sort of laboratory, in that it housed much equipment, and consequently, the astronauts spent much of the day working there. Inside, there were work and control panels for the solar telescope—a large console with television screens and other visual indicators that allowed the astronauts to monitor the telescope's operation and to take part in the selection of objects to be observed, the consequent pointing of the instrument, the experimental

procedures and the final interpretation of the images. The television system was linked with other remote stations distributed throughout the laboratory modules, and the images could be recorded on a video recorder in the multiple docking adapter. The console also tracked Skylab's attitude control.

Another system, the Earth Resources Experiment Package (EREP), consisting of sensors and cameras positioned on the side facing away from the Sun, allowed the astronauts to observe terrestrial resources. A third type of installation, called M515, was for experiments in producing specimens of new materials in zero gravity.

At one end of the module's longitudinal axis was the main docking system, with which the Apollo capsule docked when it arrived from Earth with the crew. A second, emergency docking system was situated on the side.

THE APOLLO TELESCOPE MOUNT

The most important scientific instrument onboard Skylab was the telescope, or rather observatory, which was so large and complex as to be a module itself (weighing 10 tonnes), albeit full of equipment rather than astronauts. At launch, the telescope (actually a cluster of telescopes) was placed on a truss to hold it down in front of the multiple docking adapter. Once in orbit, the truss was turned 90° to free the main docking system and

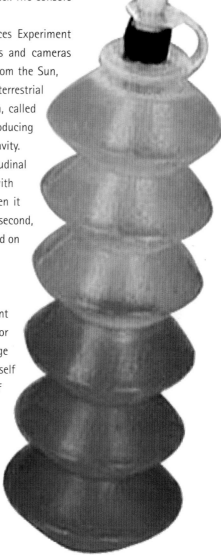

then placed crosswise to the laboratory. In this position, the observatory could deploy the four solar-cell panels from which it drew its power.

The observatory had an octagonal structure consisting of a rack 3.3 meters in diameter and 3.6 meters high to which solar panels and batteries were attached and inside of which was a container housing eight telescopes. The container was equipped with a liquid (water/methanol) refrigeration system that maintained a constant 12°C to ensure the thermal stability required by the instruments.

The instruments, designed for observing the Sun, were an X-ray spectroheliograph, an X-ray telescope, a white-light coronagraph, a spectroheliograph for extreme ultraviolet and a spectrometer-spectrohe-

liometer, also for extreme ultraviolet. In addition, there were two telescopes equipped with hydrogen-alpha filters.

The cluster of telescopes could rotate around its vertical axis by about 120°, and the axis could tilt by about 2° with respect to the rack axis. Many of the instruments recorded images using film that was recovered and replaced by the astronauts during their space walks. The majority of the telescopes were activated only when the astronauts could monitor them directly. Only three were operated automatically, even when the station was uninhabited.

ORBITAL POSITION CONTROL

Controlling its orbital position was essential for Skylab, because its eight instruments to observe the sky and its other devices for Earth-surface reconnaissance all needed to view their subjects precisely. For example, aiming at the Sun had to be done with an error of no more than 2.5 arc-seconds maintained for at least 15 minutes. To achieve this, the observation system used guide stars: the Sun and three stars (Canopus, in the constellation Carina, being preferred).

The orientation-control system was made up of three parts. First, a group of nine gyroscopes (three for each axis, plus one reserve) measured the angular rotation of each of the three axes to provide the basic elements for determining position or shifts. A digital computer collected the signals and at the same time checked that the whole thing was functioning properly. Using the guide stars and data from the gyroscopic platforms, the central computer was able to maneuver the station using the large gyroscopes or the traditional motors. Second, there were three large gyroscopic systems (Control Moment Gyro System), with a rotor 55 centimeters in diameter and weighing 65.5 kilograms that turned at 9,000 revolutions per minute. The three were mounted perpendicularly to each other, and electric motors could change their position to induce movement in the entire station, thanks to their large size. This was the main system used to move Skylab, because it was clean and did not pollute the surrounding atmosphere and impair observations with the scientific instruments. It was

Right, container with astronaut's personal effects. Below, checking mouth and teeth.

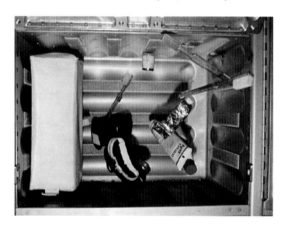

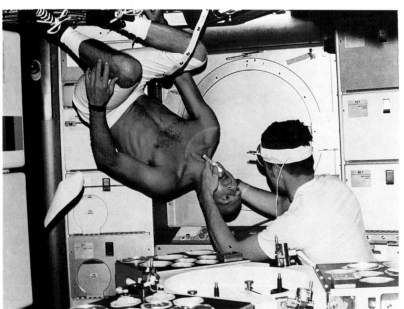

the first time that large gyroscopes had been used for stabilization. Two were sufficient for all operations, the third being redundant. All the gyroscopes—that is, those for position control and the bigger ones for attitude control—were in the Apollo Telescope Mount together with the telescope's communication equipment.

The third part of the system was a traditional propulsion system composed of two sets of three cold gas rockets (using nitrogen, a clean gas) situated on opposite sides at the back of the laboratory. They were able to give a maximum thrust of 25 kilograms. The nitrogen tanks were 22 titanium spheres located at the back of the station. However, this rocket system was considered a reserve method for attitude correction and was used only in certain phases of the mission, such as docking with the Apollo command module. When the astronauts wanted to maneuver Skylab, they typed the information for the new attitude into the computer, which compared the current position (using the small gyroscopic platforms) with the requested one and then shifted the axes of the large gyroscopes to make the desired changes.

SOLAR POWER

The power for Skylab came from two sets of solar panels: four were around the Apollo Telescope Mount, each 11 meters long, supplying 3.7 kilowatts; another two (9 meters long) were on the sides of the laboratory and generated 3.8 kilowatts. The two sets of silicon solar cells were independent but could augment each other's power, if necessary. This capability was to prove extremely valuable. When the station was in shade, nickel-cadmium batteries (8 in the laboratory and 18 smaller ones in the observatory) came into action and then recharged when the panels were again exposed to the Sun.

DEPARTURE AND ACCIDENT

On the morning of May 14, 1973, a Saturn V rocket blasted off with no delays from pad 39A at Cape Canaveral. The temperature was high, as was the level of satisfaction at finally seeing the first American space station fly. It was not yet the large station dreamed of, and experimentation would have to con-

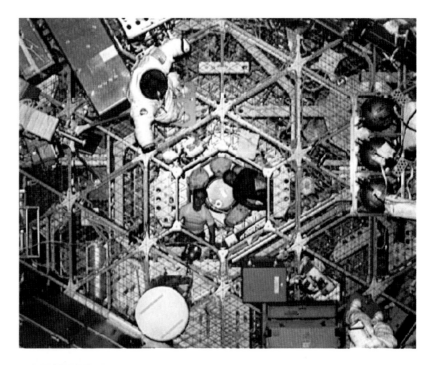

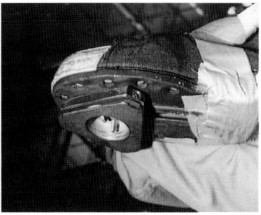

Above, the floor with triangular holes into which a protrusion on the bottom of the astronauts' shoes, left, fit to hold them in place.

tinue for a decade, but this effort truly began the process of learning about life in orbit.

However, 63 seconds after launch, when the rocket was climbing normally toward outer space, two signals were received at mission control that at first seemed rather incredible but turned out to be indications of serious trouble. One signal indicated that the antimeteorite shield had risen prematurely to its final position, and the other that the two solar-cell panels on the sides of the laboratory had opened. (Of course, both operations should have occurred only in orbit). Ten minutes after launch, the laboratory separated from the second stage of the Saturn V as planned, and eight seconds after that, it was in a circular orbit 433 kilometers above the Earth with an inclination of 50° to the equator. At this point, the station was activated and oriented so that the solar-cell panels and the telescope would always point toward the Sun.

But no signals arrived to confirm that the antimeteorite shield and the lateral solar panels had opened, and two commands sent for that purpose received no response. Meanwhile, telemetric data being transmitted to the ground stations indicated that the temperature inside Skylab was rising to alarming levels.

At this point, mission control at Houston began to

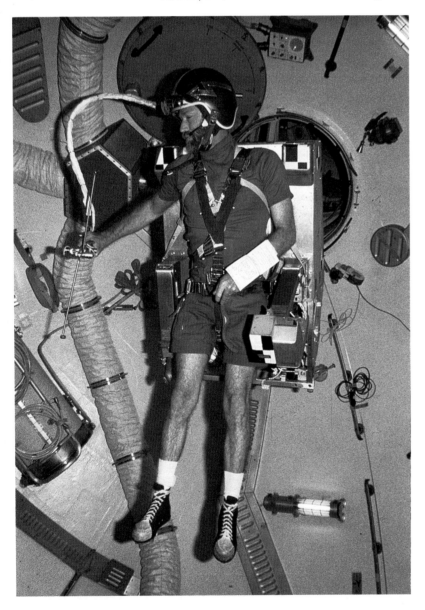

understand what had happened 63 seconds into the flight, when they had received the two surprising signals. The antimeteorite shield must have been torn away during the flight, together with one of the solar panels.

The launch of the first crew—Charles Conrad Jr., Paul J. Weitz and Joseph P. Kerwin—scheduled for the following days, was therefore put on hold while a solution was sought. In the meantime, the laboratory was positioned so that its internal temperature did not exceed emergency levels while still allowing the solar panels deployed around the telescope to be partially illuminated so as to generate at least some power (2.8 kilowatts). As the two electrical systems could be linked together when necessary, this power could also be used to run the essential parts of the laboratory.

The rise in temperature was due to loss of the antimeteorite shield's secondary function as a heat shield—without it, there were serious problems. To make matters worse, the exterior of the laboratory was covered with a thin gold sheeting that was originally intended to balance the heat exchange in the inter-wall cavity. But without the protective shield, the gold leaf, being an extremely good absorber of solar radiation, contributed to heating the station to intolerable levels.

NASA mobilized its centers, as well as the universities and corporations involved, to find an answer. The result was a simple and effective solution, which later, combined with the skills of the astronauts, overcame the problems that threatened in the initial hours after launch to make the first American space station a total failure.

From among the various suggestions, it was decided to use the airlock on the side of the telescope—designed for exposing experiments to space—to mount a sort of umbrella, a parasol made of aluminized plastic. Once opened with a mechanism controlled from inside, it would shade the gold surfaces and reflect solar radiation. The device was hastily built and put through the necessary endurance tests. The process of opening the umbrella was also tested.

Astronauts Conrad, Kerwin and Weitz on their return to Earth at the end of the first Skylab mission.

First, however, it was necessary to know what had happened inside the laboratory as a result of the high temperatures. Besides some electronic systems being exposed to dangerous degrees of heat and the organic material and drugs stored onboard likely being altered, the greatest fear was, perhaps, the possibility that the polyurethane shield (used as thermal insulation on the inside walls) was decomposing and releasing poisonous gases. Added to this was the fact that the attitude-control operations had consumed half

the nitrogen for the entire Skylab mission. Since reserves exceeded 25 percent of the quantity optimally needed, there was still an acceptable margin. As for the second solar panel, the information received suggested that it had not opened because it was blocked by fragments of the torn shield. The astronauts would free it using specially prepared tools.

THE FIRST CREW AND REPAIRS

Once the new systems had been installed in the Apollo capsule, the three astronauts of the first crew—Conrad, Kerwin and Weitz—lifted off with a Saturn I-B rocket on May 25. As the rocket soared through space, the three optimistic men sent back an encouraging message to Earth: "We can fix everything."

The commander, Conrad, was a space veteran, having flown in the two-seater capsules Gemini 5 and 11 and walked on the Moon with Apollo 12. Kerwin was a Navy doctor and pilot, as was Weitz, who had trained, however, as an aeronautical engineer. The latter two were making their space debut.

Apollo: the transport capsule

The trip from Earth to Skylab and back again was made via the same type of Apollo spacecraft that had been used on the Moon missions. It consisted of a command module capable of accommodating three astronauts as well as a service module equipped with a propulsion system and its own power reserves. Launch was from Cape Canaveral on a Saturn 1-B rocket.

Once it reached Skylab, Apollo berthed with the multiple docking module and its systems were shut down. However, it

remained ready as a rescue vehicle in case of emergency. The crew returned to the capsule at the end of their mission for reentry using the traditional sequence employed on the lunar missions: the service module separated, and the command module (protected by a heat shield) entered the atmosphere and parachuted to splashdown in the ocean. A modified Apollo capsule had been prepared for any emergency: at the base, behind the seats for the three astronauts, some fixtures had been removed

to make room for two more seats. The emergency arrangement could be set up in eight hours, and the capsule would then take off with only two pilots, leaving

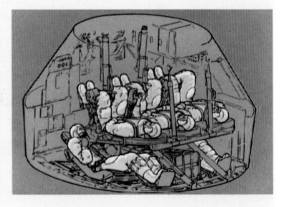

three places free for the astronauts in difficulty. Five people would therefore return. The time to arrange a rescue mission was estimated at 10 to 48 days.

The photograph shows an Apollo capsule being launched, while the drawing illustrates the modified version for five astronauts.

Eight hours after launch, as Apollo approached Skylab, a first inspection of the outside of the station was made through the portholes of the capsule, revealing that the guesses made on the ground were in fact correct. The antimeteorite shield had been torn off, together with the number-two solar panel. Number-one solar panel was blocked by pieces of metal. At this point, Apollo docked with Skylab, and the astronauts were allowed to have something to eat before facing the hardest part of the day.

The first thing to do was to free the solar panel to provide power for the laboratory. So Apollo undocked and moved toward the critical area. Weitz, dressed in his space suit, opened the door and used a long, hooked stick to try to free the debris blocking the panel from opening. The operation failed, and as orbital night was approaching, Conrad maneuvered the capsule toward a new docking, but without success. The mechanism had jammed. Once more in their space suits, the astronauts had to dismantle part of the docking probe in the depressurized cockpit. Luckily, the fault was fixed, and the union with Skylab went ahead without further difficulty.

Now the astronauts had to enter the laboratory, and the 10 days it had spent in orbit in undesirable conditions left a lot of unanswered questions. Had the systems held out? Finding out was risky business. Weitz put on a gas mask and entered the station to test the internal atmosphere for any toxic gases. The results were negative, so the entire crew went in to carry out "operation parasol."

The temperature inside was 54°C, with very low humidity. But the initial systems were activated, and the aluminized umbrella was installed at the airlock, ready to be pushed outside and opened. Conrad extended it very cautiously, and finally, the parasol opened completely, with only one bent corner.

With this new protection, the internal temperature immediately went down to a more acceptable 35°C, allowing the activation of the laboratory to continue. But at this point, the exhausted astronauts moved into the multiple docking adapter, where the temperature was a more comfortable 20°C. While they slept, the images received on the ground were used to study

the maneuvers the crew would have to carry out in a space walk to free the solar panel at last.

On June 7, Conrad and Kerwin walked around the outside with their cables and tools. Using large scissors, they cut away the fragments of metal and tried to open the panel. This was only partly successful, as the structure lifted just a little. So Conrad hung a cable between the panel and the telescope and raised the cable with his shoulder to get better leverage. The structure holding the folded solar panel sprung open at a 90° angle. Three and a half hours had been needed. At that point, the Sun warmed the panel release mechanism as expected, and the panel was finally deployed to the general satisfaction of the astronauts in orbit and everyone at NASA on the ground. The mission was rescued, and the available power rose from 4 to 7 kilowatts, guaranteeing continuation of the work of the mission.

Left, Saturn 1-B with the second Skylab crew onboard ready for launch from Cape Canaveral.

Above, Skylab mission patch.

LIFE ONBOARD SKYLAB

Life onboard Skylab was precisely regulated. The three astronauts all slept at the same time for eight hours. On waking, they had three hours for personal hygiene, breakfast and housekeeping, followed by a further three for experiments. Then there was an hour's pause for lunch and half an hour to relax, followed by three and a half more hours for experiments. At this point, the space day drew to its close with an hour to put things in order, followed by dinner and relaxation (one and a half hours) and planning for the next day. Then there was an hour's free time, finishing with personal hygiene before going to sleep, each in his own tiny compartment with a bed of multilayer cloth sacking attached to the wall, a small cupboard and a curtain

that acted as a door to give a little privacy. Such was the normal division of time, but obviously, alterations were made when necessary.

The large space laboratory was almost 50 meters long and 7 meters wide. Inside, after the initial problems were past, it was warm, comfortable and welcoming and there was enough room even to play. "Every American kid would love to jump around up here," Commander Conrad joked. The interior was lit by 78 fixed lamps and 5 portable ones, all of which could be adjusted down to zero. As the low pressure meant that the astronauts could not talk to each other from a distance, 13 communication points had been installed to form a telephone network enabling astronauts to talk from one part of the laboratory to another and communicate with mission control in Houston.

Everything needed for the three crews had been stored on Skylab. For example, the wardrobes contained 700 pieces of fireproof clothing, including 39 jackets, 64 pairs of pants, 201 T-shirts, 286 pairs of socks and 30 pairs of shoes. There were also 840 reusable flannel cloths and 420 towels for washing.

The designers had even thought about free time and had loaded card games, a dartboard, balls, books and tape recorders. However, few of these objects were used, for the greatest entertainment for the station's inhabitants proved to be looking out of the portholes at Earth. Conrad remembered, "It was fascinating to observe the curve of the Earth from 430 kilometers up."

To help the astronauts stay still in zero gravity, the technicians had prepared a simple but ingenious system: the floor consisted of a grille with triangular holes, and the outer sole of the astronauts' shoes was also triangular to fit into the grille and hold the foot in place.

THE FIRST STAY ENDS, AND TWO MORE
MISSIONS ARE LAUNCHED

Nearly four weeks went by, then Conrad, Weitz and Kerwin had to prepare for their return. For the third time, space suits were donned, and Conrad and Weitz went out to collect the film from the telescope. The suit designed for Skylab was not totally self-sufficient. An umbilical cord linked the astronaut with the spacecraft, supplying him with the oxygen he needed and controlling the climate inside the suit. This system caused some problems, but none insurmountable.

On June 22, the Apollo spacecraft was reactivated and checked to make sure everything worked properly. Once everything to be taken back to Earth was inside, the crew closed the connecting doors and Apollo de-docked from the station. A last check was made around the outside of the laboratory, and the capsule began its reentry, splashing down in the waters of the Pacific Ocean some 1,500 kilometers west of San Diego, where the aircraft carrier *Ticonderoga* was ready for the recovery.

The first three inhabitants of Skylab had spent 28 days and 50 minutes onboard, the longest period so far spent in space, demonstrating that humans could adapt to a long period without gravity. Apart from a lot of medical and biological information and data from the experiments in materials science, the astronauts brought back with them 25,000 photographs of the Sun and 7,500 images of Earth.

Skylab remained uninhabited for over a month, until July 23, 1973, when the second crew left Cape Canaveral at 7 in the morning on a Saturn I-B. The commander was Alan L. Bean, who had been a companion of Conrad on the Apollo 12 lunar mission. With him were a scientist, Owen K. Garriott—an expert in electrical engineering—and Jack Robert Lousma, a Navy pilot and aeronautical engineer. Both were having their first experience of space.

Their launch had been scheduled for three weeks later, but the rapid degrading of the gyroscopes that controlled Skylab's attitude and the deterioration of the parasol opened during the first mission had made it necessary to move the launch forward. Onboard were the new gyroscopes, a new parasol and other replacements, food reserves to prolong the stay by 3 days over the 56 days previously planned, as well as data recorders and an unusual population: two freshwater fish, six mice, two spiders named Arabella and Anita and 50 freshwater-fish eggs.

As Apollo approached to about 700 kilometers, the crew saw Skylab appear out of the dark, thanks to the laboratory's positioning lights that would also aid in docking. While this maneuver was being completed, Commander Bean noticed a cloud of sparkling particles around the spacecraft. An instrument check told him that there was a liquid-oxygen fuel leak, and he shut down one of the four rocket engines. This caused no problems; Apollo docked perfectly with the station, and immediately, the materials brought from Earth were transferred. However, six days later, Houston mission control noted a decrease in pressure in another of Apollo's engines and shut it down, too, as a precaution.

That decision created a risky situation—without the required engine power, it would have been impossible to use Apollo for reentry at the end of the mission, thus necessitating the launch of a rescue capsule to bring the crew home. After a tense few days examining the data, however, the capacity of the second damaged engine was judged to be adequate, the alarm was called off, and the mission proceeded.

In the first three days of their stay, the crew suffered from the traditional "space sickness," with nausea and

The second Skylab crew (from left to right): Owen K. Garriott, Jack R. Lousma and Alan L. Bean.

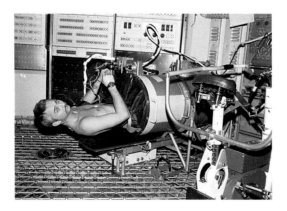

Above, astronaut Garriott in Skylab seated at the controls for the solar telescope. Below, Garriott undergoing decompression of the lower limbs.

vomiting, which naturally reduced their capacity for work. But at the end of the first week, all three were well again and began preparing for the first important extravehicular activity.

In a 6-hour, 31-minute space walk on August 6, 1973, Garriott and Lousma set up a second parasol-type sun cover, made of silicon rubber, placed over the old one, which was now damaged by ultraviolet radi-

ation and had been lowered. The new protective sheet was 7.5 meters long and 7 meters wide. It had actually been brought up by the first crew but had been stored until it was decided when to use it. The astronauts mounted a light, V-shaped tubular structure at the exit and attached the two lower corners of the parasol to the two open ends. The two upper corners were each tied to a cable that was then stretched and fixed to the base of the telescope. The whole thing was the largest structure ever assembled in space, with an overall length of 25 meters. The astronauts also installed a meteorite counter and other instruments.

Two more space walks followed on August 24, by Garriott and Lousma, and on September 2, by Bean and Garriott, to collect and replace the film in the telescope and to install other equipment. One of the new experiences of the second crew was testing a rudimentary rocket pack (to be used for future extra-vehicular activities) under the cavernous cupola of the laboratory. The test proved valuable but was also quite difficult.

After eight weeks, on September 25, the second Skylab mission came to an end with a splashdown 300 kilometers southwest of San Diego, where the *U.S.S. New Orleans* picked up the crew and their capsule.

A long interval of nearly two months passed before the third and last mission lifted off on November 16. This time, the protagonists were three new astronauts all going into space for the first time: Gerald P. Carr and William R. Pogue, both pilots and engineers, and Edward G. Gibson, a physicist and engineer. Their launch followed a now-familiar pattern: the Saturn I-B ignited its engines at Cape Canaveral at 9 in the morning. Then came the scheduled change in trajectory, from a low orbit of 150 to 227 kilometers to the higher one, where Skylab was waiting at 430 kilometers. Then came the final docking, with some problems. Nothing can be taken for granted in space travel. Two attempts failed, and only the third allowed a stable union between the two vehicles and the beginning of work.

Not immediately, however, as the first days were spoiled by the usual space sickness. Mishaps continued, this time caused by the astronauts and some surprise decisions taken in the last weeks before launch. As this was the last Skylab mission, the flight controllers had added some in-orbit activities that made the days more intense than they had been previously. The crew protested to Houston mission control that they could not fulfill such a busy schedule. Indeed, having to do too many tasks too quickly caused errors in some experiments, while others proved difficult because the crew had not had enough time to study them before launch. Everything generated heated argument between Skylab and Houston, leading finally to an increase in the astronauts' leisure time and a modification of the program. It was the first time something of this sort had happened.

The last mission was made even more turbulent by a breakdown in one of the three large gyroscopes used to control the station's attitude. Two were enough to carry out the operations, but as the problem was caused by a lack of lubrication after so many

months of use, it was feared that the others would have the same trouble. In fact, a second gyroscope soon demonstrated similar anomalies. Although these were not quite as serious, the astronauts were forced to use the nitrogen propulsion system to control and correct the laboratory's position.

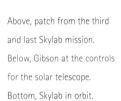

Meanwhile, the planned experiments went ahead, the rocket pack was tried out again, and four space walks were made for the usual task of collecting and replacing films, adjusting external apparatus or installing new ones. During one of these walks, on December 25, 1973, astronauts Carr and, especially, Pogue remained outside for seven hours and three minutes to photograph Comet Kohoutek with an ultraviolet camera. New Year's came, and the last space walk was made on February 3, 1974. It lasted a little more than five hours and was dedicated to collecting all the apparatus and samples exposed to the vacuum of space, as well as the last films from the telescope. And so, the longest space mission by either the United States or the USSR drew to its close: 84 days in orbit, a record that would only be broken by Russian cosmonauts four years later.

Above, patch from the third and last Skylab mission. Below, Gibson at the controls for the solar telescope. Bottom, Skylab in orbit.

The three Skylab crew members gathered a few items, like film, tools and cables, and put them in sacks in the docking module so that they would be accessible to a future crew visiting the station. On February 8, Carr, Gibson and Pogue entered the Apollo capsule, de-docked from Skylab and, after circling it for a final check, headed for Earth with the final small scare of having one of the two control systems out of order, thanks to a probable breakdown. However, splashdown went ahead without difficulty early in the morning, 289 kilometers off San Diego, as usual, where the *New Orleans* was ready to welcome the last inhabitants of America's first space station.

171 DAYS AROUND THE EARTH

The results of Skylab's 171 days of operation satisfied NASA, the engineers who had designed it and the scientists who had had the chance to conduct previously impossible experiments. A total of 117,047 photographs had been taken of the Sun, showing extraordinary details of the solar surface and phenomena never seen before. The international astronomy community now had access to an enormous quantity of data. Earth had also been the subject of repeated observations resulting in 46,146 photographs. Added to these were the astrophysical and solar observations and the hundreds of hours of experimentation in materials science, biology and space medicine. Above all, the study of the adaptation of the astronauts' bodies to long periods without gravity met one of the most important objectives of the Skylab program—that is, the investigation of how humans could live in space safely and for a long time.

But with Skylab, the astronauts had also demonstrated another determining factor for the future of human exploration of space that had not been foreseen in the program: the role of humans in technical problems and how their intervention could resolve them, transforming a failed mission—like Skylab appeared to be when it reached orbit after the accident—into an enormous success, thanks precisely to the repairs made by the astronauts.

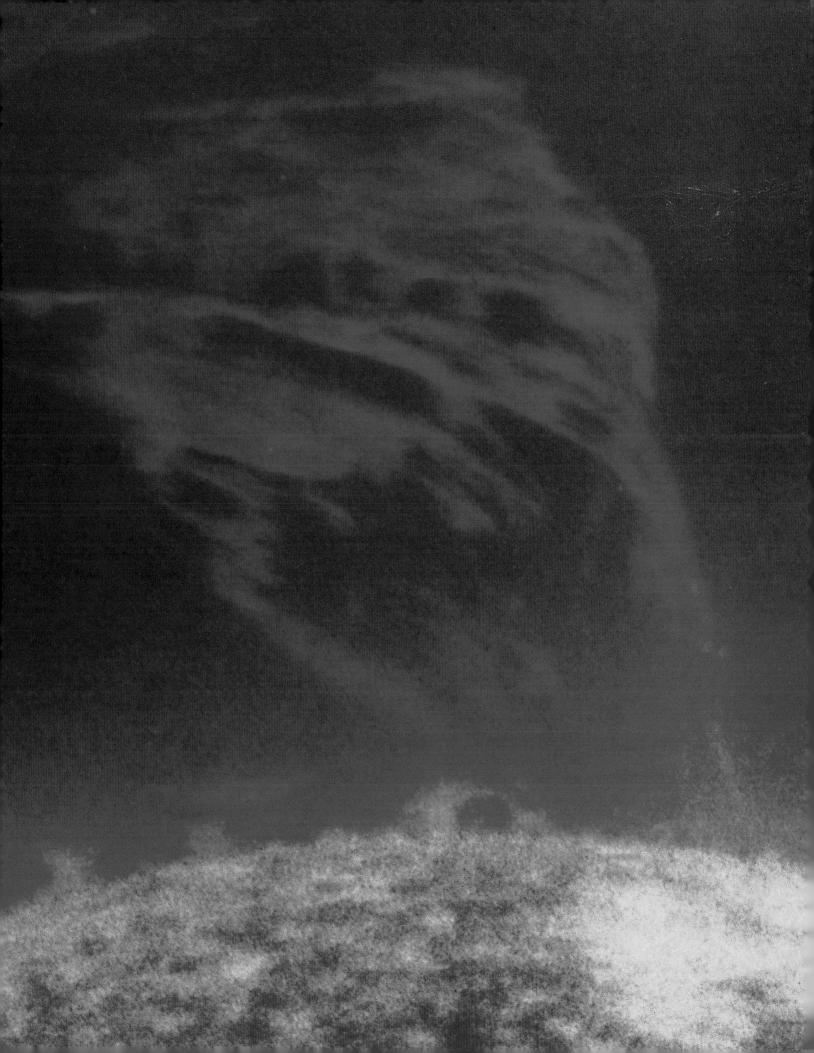

Skylab with the second Sun shade placed over the first protective umbrella.

cells received the Sun's radiation, the laboratory still managed to send telemetric data with updates on its condition. However, the situation was rather grave; two of the three gyroscopes for attitude control were practically unusable because part of the cooling system was not working, and various other systems were now at the end of their operational life.

According to the severest critics, Skylab, which had cost $2.5 billion, was now a wrecked ship, and there was nothing else to do but watch it fall to Earth and disintegrate in the atmosphere. According to calculations taking account of solar activity, the laboratory could remain in orbit until 1983. Unfortunately, the Sun went through a period of maximum activity at the end of the 1970s that was more intense than predicted and caused an increase in the density of the Earth's atmosphere. The consequences were inevitable: as it met with greater resistance in its orbit, Skylab slowed down and gravity had a greater effect, lowering its orbit more quickly and shortening its life.

The possibility was investigated of using the space shuttle (then under construction) to take a propulsion system (Teleoperator Retrieval System) into orbit and attach it to the laboratory to raise its altitude consid-

THE END OF SKYLAB

After the last crew had abandoned Skylab, some people at NASA wanted to find a way to continue using it. It had stabilized with the larger part facing down and the docking module facing up. Its systems were almost completely exhausted, but when its solar

Skylab in the Space and Aeronautics Museum in Washington.

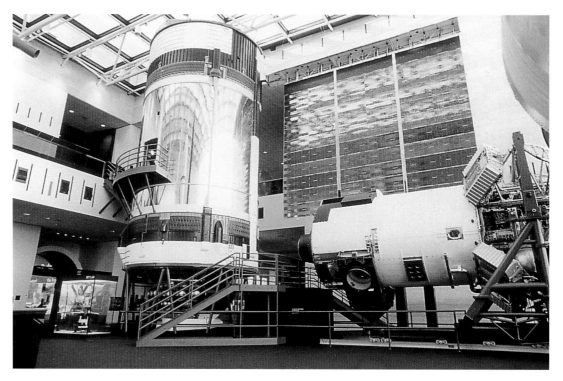

erably and make it safe for a number of years more. Then, in January 1978, the Soviet spy satellite Cosmos 954, carrying a nuclear reactor, fell in Canada, and worries about Skylab's anticipated reentry speeded up the construction of a rescue system.

However, it became clear that the time needed to complete construction of the shuttle would not coincide with Skylab's needs. The idea was therefore abandoned, and preparations were made to control Skylab's reentry as much as possible so that the impact would not be dangerous.

On July 11, 1979, Skylab disintegrated in the sky above the west coast of Australia, in an area south-east of Perth, causing an alarm that mobilized five continents, amid conflicting figures coming from NASA and the Pentagon. All the predictions had indicated that Skylab would fall in the Pacific, but some uncontrolled movements by the station in the last phases of reentry shifted the point of its impact with the atmosphere.

Several pieces landed on Australian soil and were subsequently put on show in certain Australian cities. A fragment several meters long can be seen in the museum of the Alabama Space and Rocket Center in Huntsville, near NASA's Marshall center, where Skylab was born.

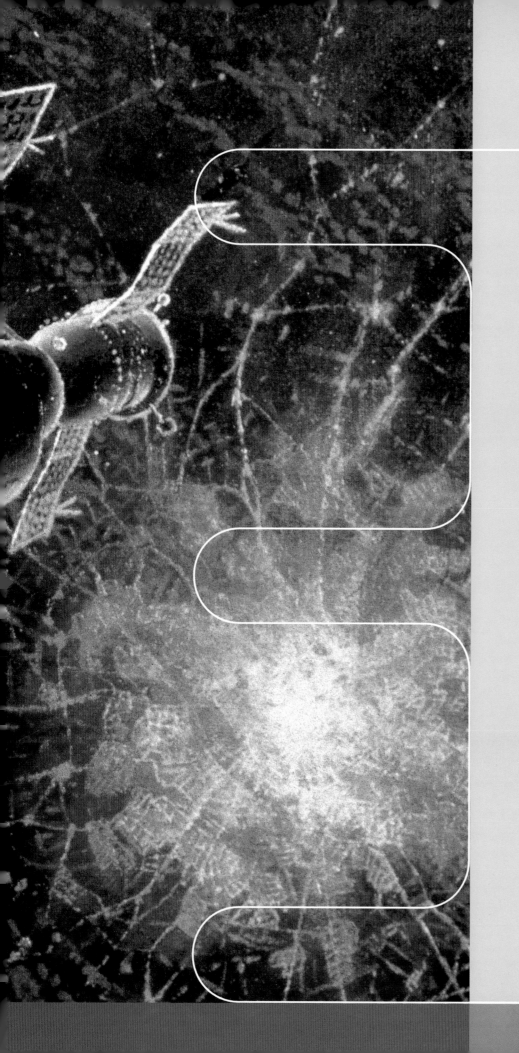

The Russian Salyut stations

Seven military
and civilian
laboratories

The Russian Salyut stations

In his diary in 1962, Sergei Pavlovich Korolev, the "father" of the first Russian space rocket and of Sputnik, wrote, "The time has come to consider Tsiolkovsky's *Oranzhereya* [Greenhouse] project," recalling a space-station design envisaged by the great pioneer at the beginning of the century.

While the 1960s marked the idea of a manned space station's passage from dream to reality in both the United States and the Soviet Union, for Korolev, it had already been part of a plan presented to Soviet armaments minister Dmitri Ustinov on May 26, 1954. It consisted of four steps: development of a 2-to-3-tonne scientific satellite, a recoverable satellite, a one- or two-person capsule and "an orbiting station having regular links with Earth."

By 1962, the first three objectives had already been met, and all that remained was to work on the station. To that end, in March 1962, Korolev prepared a report entitled "Complex for Assembly of Space Vehicles in Artificial Earth Satellite Orbit (the Soyuz)," which described a small station composed of modules launched separately by the R-7 rocket used to launch Sputnik and Yuri Gagarin. The modules would be joined in space, and three astronauts would go to the base with a vehicle called *Seyver* (North). Some of the modules would be used for living quarters, and others for working with scientific instruments.

Three years later, Korolev proposed a new, larger, 90-tonne station to be launched with the N-1 rocket developed for a manned flight to the Moon and equipped with four docking systems for the *Soyuz* (Union) capsules.

Like the Americans, the Russian military viewed space as an advanced reconnaissance post, and further incentive to proceed in this direction came from the United States in December 1963, when Defense Secretary McNamara announced termination of the Dyna-Soar mini-shuttle program and the beginning of the military MOL (Manned Orbital Laboratory) project.

Preceding pages: an illustration of the Russian station Salyut 6.

Korolev's design for the *Seyver* (North) station.

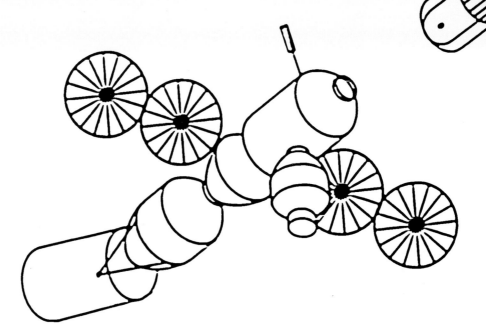

RUSSIA'S MILITARY SPACE-BASE PROJECT

The Russians were not slow to respond, and within a few months, Vladimir Chelomei, the USSR's director of military missile construction, drew up a program that looked like a carbon copy of the United States' MOL. Chelomei was Korolev's worst enemy, because he had the support of the military; one of the people working for him was engineer Sergei Khrushchev, son of the famous premier. This position of special favor naturally came to an end when Leonid Brezhnev replaced Khrushchev as premier in October 1964. But Chelomei found a new ally almost immediately in Marshal Andrei A. Grechko, the powerful defense minister.

This situation gave rise to the development of two space-station programs: the military *Almaz* (Diamond) and the civilian *Salyut* (Salute/Greetings). However, the veil of secrecy hanging over affairs behind the Iron Curtain gave out only the one name, Salyut, for both the military and the civilian versions, thereby helping to confound the issue and guarantee a cover.

In October 1964, Chelomei drafted plans for a laboratory and a truncated-cone-shaped capsule. Both the name given to the capsule—Merkur—and its shape were highly reminiscent of America's Mercury space-craft.

Almaz consisted of two cylinders, one with a diameter of 4.15 meters and one of 3 meters, connected by a truncated-cone-shaped midsection. The Merkur capsule would dock with the narrowest end. The capsule could carry two or three cosmonauts and was equipped with an emergency rescue system that could be used during the early phases of a launch. The crew stayed inside the capsule during launch or reentry, but once in orbit, they moved into the laboratory via a connecting door located in the heat shield at the bottom end of Merkur (just like the Pentagon's design for attaching a Gemini capsule to the MOL).

As Chelomei told Road Z. Sagdeev, director of the Moscow Institute of Space Physics (IKI), Khrushchev was keenly interested in Russian aircraft carriers, and the proposal for a manned orbital base equipped with observation systems, including radar, was immediately accepted by the Kremlin. The station was not going to be very big, because, as Chelomei pointed out, "the

maximum diameter for transporting components on the railway was exactly 4 meters."

A detailed Almaz program was officially presented to the VPK military-industrial commission in 1967, and it was approved. It described a two-part structure: one section was the laboratory, and the other the TKS supply vehicle, which would not only transport materials and fuel but could also be attached to the laboratory and used like a tug to shift the station's

Above, Sergei Korolev.
Below, Yuri Gagarin, the
first cosmonaut.

Facing page, two cutaway
views of the first Soviet
station, Salyut 1, with a
Soyuz capsule attached:
1) radio antenna for the
rendezvous system
2) solar panels
3) telemetry antenna
4) portholes
5) Orion telescope
6) air-renewal system
7) TV camera
8) camera
9) instruments for biological
research
10) refrigerator
11) berths
12) water tanks
13) waste containers
14) attitude-control thrusters
15) fuel tanks
16) bathroom
17) micrometeorite sensors

orbit. Both station elements had a Merkur capsule able to accommodate a load of 360 kilograms, or three cosmonauts.

Almaz was to be fitted with reconnaissance radar to obtain images of the Earth's surface, as well as a video camera and two small capsules to return the film to Earth. It was also planned to have an antiaircraft cannon for defense in the event of an American attack. Everything was to be built in the Khrunichev factory.

Almaz and the TKS tug would leave separately, each on a UR-500K rocket (the equivalent of a modern-day Proton), and would then be connected at their aft ends, where a docking system with a door provided passage for the cosmonauts. The first flight of the Almaz/TKS/Merkur space-station assembly was planned for 1968.

Meanwhile, Korolev and Chelomei, respectively heading project centers known as OKB-1 and OKB-52, were involved in a bitter competition to develop a manned lunar program. Chelomei's plan was for a circumnavigation of the Moon and return to Earth, while Korolev wanted a landing. To that end, he was constructing the large N-1 rocket and adapting a new capsule called Soyuz, which he was developing, to

have a vehicle with greater capabilities than the original, primitive Vostok and Voskhod capsules.

DISASTER STRIKES THE FIRST SOYUZ CAPSULE

In any event, the first manned flight of the Soyuz capsule in April 1967 ended in disaster: on reentry, the parachute failed to open, and the tremendous impact with the ground caused the death of the one cosmonaut onboard, Vladimir Komarov. Korolev did not witness this sad outcome, however, as on January 16, 1966, after a brief hospitalization, he died during surgery to remove a colon tumor.

Vasily P. Mishin, for many years Korolev's right-hand man, was nominated as head of what would remain known as the Korolev Center. After the necessary modifications, Soyuz flew again, this time successfully.

On the other hand, things did not go well for the N-1 lunar rocket. Its first two launches in February and July 1969 ended in explosions shortly after lift-off. Then in July, the Americans landed on the Moon with Apollo 11, and the USSR realized it had totally lost the great challenge that had been kept rigorously secret and officially denied.

Stations in science fiction films: *Solaris*

In 1972, Andrei Tarkovsky's film Solaris was released in the USSR. It is the story of Kris Kelvin, a scientist sent to investigate a space station orbiting the planet Solaris. He discovers radiation capable of instilling anguish and desire in the crew.

The film was based on a book by Stanislaw Lem and was publicized as the Soviet answer to 2001, A Space Odyssey.

THE BIRTH OF THE SALYUT 1 STATION

Chelomei's plans for a military station proceeded slowly and with numerous difficulties; and in February 1970, the ministry to which he answered (the ministry for machine construction) ordered that the program be transferred to the rival group directed by Mishin. The mandate was to review the Almaz project with an eye to using Soyuz capsules instead of Merkur, which remained on the drawing board.

Meanwhile, the United States was working on Skylab, a fact that accelerated the Russians' desire and efforts to be first. It was the Korolev Center's chance for revenge, and Mishin, well aware of that, took up the challenge and immediately introduced a series of modifications. One of these was replacing the planned two rear engines with the engine block previously used on Soyuz and literally just tacked onto the new vehicle. This meant, however, that the docking system for the capsules could not be placed at the rear, and it was moved forward to the nose. On the engine block's sides were two solar panels, also of the type used on Soyuz. To save time, the alterations were made at the Khrunichev factory, where both Korolev's Soyuz and Chelomei's Almaz base were built.

And so, the first Soviet orbiting laboratory, Salyut 1, was born. Its main objective was to test the systems and technologies necessary to carry out civilian and military operations during a long manned sojourn in space.

The base was composed of three elements:

1) Docking and transfer compartment. Located in the forward section, this was a cylinder ending in a truncated cone containing the docking system. The whole assembly was 3 meters long and 3 meters in diameter. A door separated this compartment from the rest of the station. Inside were the environmental control panels and some research instruments, while outside, there were two solar-cell panels, a TV camera and the antenna needed during rendezvous, plus a light to help with manual docking when necessary.

The exterior also held the Orion 1 optical telescope built by the Armenian Academy of Sciences and controlled from inside the spacecraft. Through an airlock, a robotic arm was used to change the spectrograph cassettes. A second instrument used for astronomy was the Anna III gamma-ray telescope built by the Moscow Institute of Engineering and Physics. It was named after the 8-year-old daughter of one of the scientists, because, like her, the researchers had much to learn (in this case, about gamma rays). The compartment also had a door for extravehicular space walks.

2) Work and rest compartment. This 9.1-meter-long section was the heart of the station. It was composed of two cylinders with different diameters. The first, linked to the docking compartment, was 3.8

18) main corridor

19) worktable

20) control center

21) pressurization system tanks

22) collimator

23) Soyuz capsule engines

Station dimensions:

23 meters long with Soyuz capsule (16 meters without), widest diameter 4.15 meters, weight (with Soyuz) 25 tonnes.

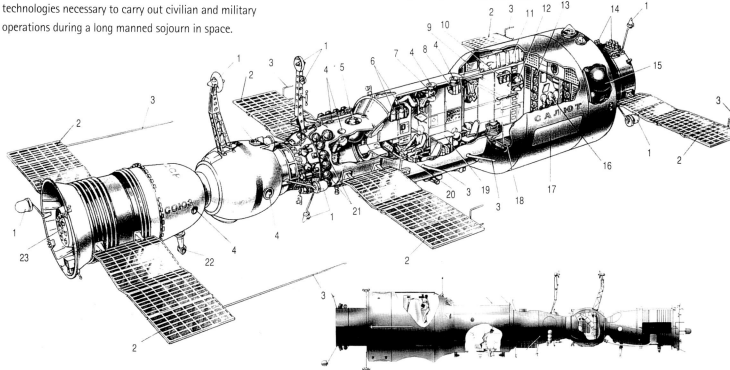

meters long and 2.9 meters wide, while the second was 4.1 meters long with a diameter of 4.15 meters. A 1.2-meter-long truncated cone connected the two cylinders, which had semicircular ends.

The largest area was dominated by a cumbersome cone rising from the floor and housing equipment used to observe the sky and the Earth. This space also held water and food reserves, two refrigerators, exercise equipment and emergency medical supplies. One wall contained the hygiene unit, with a ventilation system and a system for washing the walls. Sleeping bags were hung on the walls. Alternatively, if the cosmonauts preferred, there were the berths in the Soyuz capsule.

The smaller area housed the station's control systems (the main panel was derived from the one in the Soyuz capsule), including the large gyroscopes for attitude changes. A table with a portable water container was also provided. Here, the cosmonauts ate and passed the time with a cassette recorder, a drafting table and a small library.

The interior was lit by 27 lamps distributed throughout. To aid in orientation, each wall had been painted a different color (light or dark gray, apple green, pale yellow). The cosmonauts breathed air composed of oxygen and nitrogen, and a forced-ventilation system was used, because without gravity, there was no spontaneous circulation. Excess heat was expelled into space by radiators. Heat given off by all the machines and equipment in the forward area was absorbed by the atmosphere, which was passed first through heat exchangers and then through regenerating cartridges that finally returned it to the original environment. To guard against any nasty surprises, the external walls of the habitation zone were protected by antimeteorite and heat shields.

3) Engine compartment. Attached to the rear, this section was 2.17 meters long with a diameter of 2.2 meters. It was unpressurized and inaccessible to the cosmonauts. Two silicon solar-cell panels similar to those installed in the forward compartment extended from its sides. The propulsion system was derived from the one used on Soyuz (KDU-35) and consisted of the main engine, with a thrust of 417 kilograms, plus a reserve (with two exhaust nozzles) yielding an almost equal thrust of 411 kilograms. To these were

added four smaller engines, each with 10 kilograms of thrust, for attitude control, and around these were four sets of small rockets to control pitch and yaw. Another two pairs of tiny rockets controlled roll. This compartment also contained the fuel tanks.

DEPARTURE BUT NO DOCKING

The first station was ready in 12 months, and on April 19, 1971, a Proton rocket blasted off from Baikonur to take it into orbit. Until just before launch, the station's name was *Zarya* (Red Sky), but it was renamed *Salyut* (Salute/Greetings) to celebrate the 10th anniversary of the launch of Yuri Gagarin, the first

Drawing, right, and photograph, below, of the docking system connecting a Soyuz capsule (male) and the station (female).

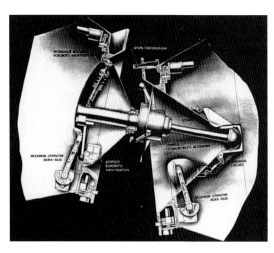

man in space. The rocket placed it in an elliptical orbit 176 to 211 kilometers high. Ground control then ordered ignition of the station's engines, taking it up to a higher orbit of 200 to 222 kilometers with an inclination of 51.6° to the equator. It took 88.5 minutes to circle the globe. The next day, Premier Brezhnev proudly announced that the new launch was "an important step in the conquest of space."

While still empty, the station was stabilized by rotating slowly (3° a second) on its longitudinal axis. The same was done when the cosmonauts onboard did not have to make observations requiring a stable base. At the same time, the position-control system kept the orbiting assembly's solar panels pointing toward the Sun to generate power.

Salyut 1 was not alone for long. On April 23, the Soyuz 10 capsule left from Baikonur carrying Vladimir Shatalov, Alexei Yeliseyev and Nikolai Rukavishnikov (crew code-named *Granit* [Granite]). The first two cosmonauts had been selected because they had achieved the first docking in orbit between the space capsules Soyuz 4 and Soyuz 5 in January 1969.

After 22 orbits of the planet, Soyuz 10, which was being tracked by the Yevpatoriya station in the Crimea, drew near to the station. The automatic system took it to 180 meters from the target, then Shatalov manually controlled the last phases of docking. It was April 24.

However, the new docking system began to act up. Instrument checks showed that the electrical connections were faulty; pressure refused to equalize between the two cabins, and the opening mechanism for the entrance into Salyut was jammed. Docking, in effect, had been incomplete and inadequate. It was therefore impossible to enter the station, and after 5.5 hours, Soyuz 10 separated, spending an hour conducting detailed photographic reconnaissance around the base. The news agency Tass, however, proclaimed that the capsule "had completed the planned experiments." Sixteen hours later, the capsule made an anomalous landing in the dark 120 kilometers north of Karaganda in Kazakhstan, coming to rest only 50 meters from a lake. "A gust of wind saved us," the cosmonauts said later.

SECOND FLIGHT, SECOND DISASTER

Less than two months later, Soyuz 11 took off from Baikonur on June 6 with a modified docking system. According to Russian regulations, the crew should have been Aleksei Leonov, Valery Kubasov and Piotr Kolodin, the reserve crew for the first flight. But two days before launch, Kubasov came down with a throat infection, and the entire trio were replaced by their reserves: Commander Georgi Dobrovolsky, mission engineer Viktor Patsayev (both on their first flight) and scientist Vladislav Volkov, a veteran of Soyuz 7. They were given the code name *Yantar* (Amber).

On June 7, after the usual race through space, Soyuz reached the station, which was in an elliptical orbit 177 to 209 kilometers high. Docking was manually controlled by Dobrovolsky, and this time, the docking system functioned correctly. Two hours and fifty minutes later, the three cosmonauts made the first entrance to Salyut 1. For two days, they worked at activating its various sections, raising its orbit and turning on the scientific instruments. They were enthusiastic about the experience. "It's as if the station were endless," Dobrovolsky said, happily flying around in zero gravity.

The three inhabitants of the orbital base spent 20 days on various activities: astronomical observation with the Orion 1 and Anna III telescopes; photographic studies of the Earth's continents and oceans; biological experiments with fertilized frog's eggs, fruit flies (drosophila) and unicellular green algae (chlorella, used to regenerate the air onboard); and studies of plant growth in zero gravity. Moreover, the first important experiments in space medicine were done, monitoring the functioning of the human body during long-duration space flight.

To keep the cosmonauts in good physical shape, they had 40 minutes of exercise every morning after breakfast and another 80 minutes distributed throughout the day. Psychologists also kept a continuous watch on the crew through radio and television links and noted how body language (for example, gestures and facial expressions) was completely altered by conditions of weightlessness. But they made an error: listening to the slightly changed voice tones, they attributed this to an excess of stress, whereas in fact, the anomaly was caused by the space environment.

On June 18, the routine aboard Salyut 1 was broken when the three men discovered a burn in one of the electrical wires. They were considerably alarmed and immediately asked mission control to program their return and the reactivation of the Soyuz capsule, where they took refuge. However, after it was determined that there was no real danger, the crew went back into the station and went on with their work. It was clear to mission controllers, however, that their spirit had been sapped, and a decision was made to end the mission one week thereafter.

On June 29, the crew collected the film and experimental specimens and put them in Soyuz 11. Then they de-docked from the station and began reentry maneuvers. It was the 23rd day of the mission, when fatally, a fault in a valve let out all the air in the capsule and the cosmonauts died because they were not wearing their pressurized space suits.

MISSIONS ON HOLD AND THE END OF SALYUT 1

So the first and last stay on the first space station, Salyut 1, ended in tragedy. The accident obviously put a halt to all further flights while the Soyuz systems were being redesigned. However, this required more time than the planned operational life span of the

Cosmonauts Yuri Romanenko (right) and Georgi Grechko inside a Soyuz capsule.

Above, the Semiorka rocket used to launch the Soyuz capsules.
Facing page, cosmonaut Alexander Ivanchenkov inside Salyut 6.

station, which continued to be tracked from the ground, more for the sake of learning how to manage a large orbiting vehicle than for anything else. Indeed, when the fuel available to maintain its position was almost exhausted, the station was brought to a splashdown in the Pacific Ocean on October 1, 1970, following a controlled reentry trajectory.

In the meantime, after the accident that so tragically marked the debut of the first orbital base, the leader of the cosmonaut corps, General Nikolai Kamanin, was replaced by General Vladimir Shatalov, who had been the hero of the first docking between two capsules in January 1969. In the following months, while Soyuz was being redesigned, a new station, Salyut 2, was prepared, incorporating some alterations suggested by the first experience.

TWO MORE FAILURES: COSMOS-SALYUT AND THE MILITARY STATION ALMAZ/SALYUT 2

Two years after the disaster, another Soyuz was on the launchpad, ready to leave with no cosmonauts aboard, as a test of the improvements that had been made. The test was kept secret, and the launch took place from Baikonur on June 26, 1972, under the generic name Cosmos 496 (a name used for a varied series of space vehicles, including scientific and military satellites). On July 2, the capsule reentered correctly. Thus, the space-station program could begin again.

A few weeks later, on July 29, a Proton rocket blasted off from Baikonur with a new Salyut station that had been the reserve used for ground tests of Salyut 1. However, the rocket's second stage shut down pre-

maturely, and the third stage did not have the capacity to compensate for the loss of speed. Therefore, the station could not enter orbit and fell back into the Pacific Ocean.

At the time, nothing was said about the failure, and everything passed off in silence. This accident, which followed a failure in the preceding year of the second flight of the N-1 lunar rocket, whose construction was also directed by the group under Mishin (Korolev's successor), provoked a change in the balance of space power in Moscow in favor of the rivals. It was no coincidence, therefore, that Chelomei's team now made a comeback after being supplanted from work on the military station which, although its construction had started first, had had no success due to the many delays in its development.

The year 1972 also saw the start of training for the first group of 15 military cosmonauts (7 commanders and 8 flight engineers) under the direction of General Gherman Titov and Colonel Yevgheni Khrunov. So the first military space station was built in the workshops of Khrunichev according to the original ideas of its designer Chelomei.

While the central structure with its two areas of differing diameter remained the same, some systems had been changed. The compartment with the mechanism for docking with a space capsule was moved from the front to the rear, which altered the propulsion system. The more powerful engines were placed on the sides of the compartment, and some of the small thrusters for attitude control were moved around the narrower central section. Two solar-cell panels also protruded from the sides of the docking mechanism. Instead of the Merkur capsule, which Chelomei had wanted, the link with Earth was provided by Soyuz capsules.

Quite a few improvements over Salyut 1 were made, including a new digital control system, energy supply, data telemetry and general radio apparatus, as well as refinements to the environmental system that ensured life onboard. For example, a Priboi water-regeneration system was used for the first time. Cameras installed for reconnaissance of the continents and oceans provided high-resolution images that were sent to Earth in small, retrievable capsules.

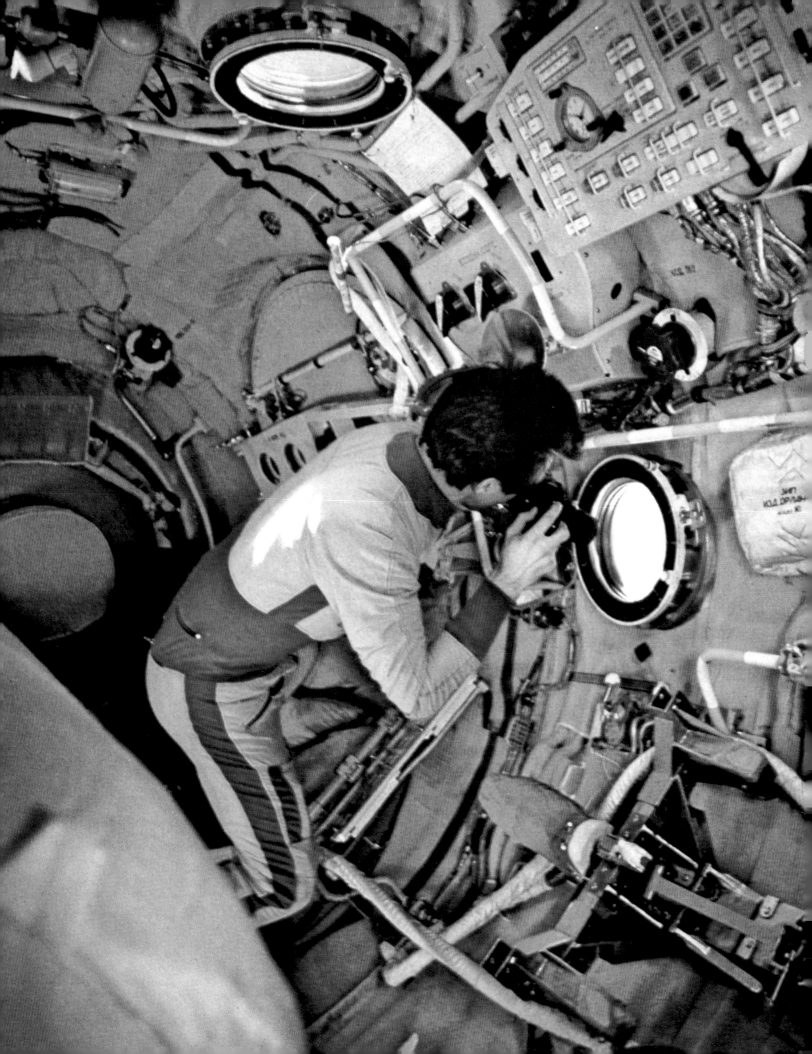

While the military called the new station Almaz 1, the official name used as a cover was Salyut 2, and it was under this name that it took off from Baikonur on April 3, 1973, much to Chelomei's chagrin, as he saw his "baby" christened with the same name as his rival Mishin's. But military rationale prevailed.

Everything took place in absolute secrecy, but the use of radio frequencies typical for military satellites and the low orbit (207 to 248 kilometers, with an inclination of 51.6° to the equator) led the West to guess at a utilization for defense needs.

In the days following the launch, there was talk of crews prepared to man the new station, which, in the meantime, had moved to a higher orbit that seemed to confirm readiness for the arrival of cosmonauts. But on April 14, Almaz/Salyut 2 suffered serious damage after another shift to higher orbit. The damage involved the attitude-control system, and the station began to spin out of control. Catastrophe followed, as the orbital base broke up and disintegrated in the atmosphere on May 28, 1973. In 1992, cosmonaut Mikhail Lisun, scheduled as a reserve for a planned Soyuz 24 flight to a hypothetical Almaz 5, claimed that the loss of Salyut 2 was due to an electrical fire followed by depressurization.

COSMOS 557 (THE SALYUT THAT NEVER WAS) AND THE NEW SOYUZ CAPSULES

Meanwhile, on May 11, 1973, a Proton vehicle had taken another Salyut station into orbit. This one, however, was a civilian version, from Mishin's Korolev Center, and was therefore of a different design from the failed Almaz 1.

But when the flight controllers at Yevpatoriya transmitted the commands to move the station from its initial orbit to the desired position, the ion-powered attitude-control system failed, igniting the engines incorrectly and using up nearly all of the available fuel. It was immediately clear that the new orbital base (launched only a few days before the launch of the first crew for the American Skylab station, which was already orbiting Earth) was to be the victim of an unhappy fate.

In fact, it disintegrated in the atmosphere a few days later, on May 22. To conceal what had happened, it was never given the name of Salyut and was officially known only as Cosmos 557.

Still, the redesign and improvement of the Soyuz capsule had proceeded, including the introduction of an important element: a plan to ensure the cosmonauts' survival in the case of vehicle breakdown by having them wear pressurized suits for both launch and reentry. Of course, for the suits to work properly, their life-support systems had to weigh quite a lot (about 100 kilograms), and since the acceptable payload of Soyuz could not be exceeded, it was necessary to reduce the crew to two cosmonauts.

1974: THE FALL OF MISHIN AND THE DEPARTURE OF ALMAZ 2/SALYUT 3, THE FIRST SUCCESSFUL MILITARY STATION

In February 1974, the first U.S. experiment with an orbital base ended with the return of Skylab's last crew, and it would take them a decade to take up the subject again. In the Soviet Union, on the other hand, interest in space stations was on the increase among both civilians and the military.

The year 1974 was a year to remember in the Soviet Union for both good and bad reasons. Premier Brezhnev signed a nuclear-weapons-control pact in June during the second visit of American President Nixon to the USSR. (Nixon was later impeached following the Watergate scandal, and Gerald R. Ford succeeded him.) Author Alexander Solzhenitsyn was expelled from the country after publication of *The Gulag Archipelago*.

It was also a year of radical change in the Soviet space program. Vasily Mishin, Korolev's successor for

The Almaz 2/Salyut 3 military station.

The Soyuz and Progress capsules and the Cosmos 1443 module

Three types of vehicle were used for transport to the Salyut stations. Soyuz T (an improved version of the earlier Soyuz) could accommodate two or three people. It had three components: a 3-tonne module for the cosmonauts' return to Earth, an orbiting module for various instruments and a service module with engines, top left, that was abandoned before reentry. Power onboard was supplied by two solar-cell panels with a wingspan of 10.6 meters. The complete Soyuz T assembly was 6.98 meters long with a diameter of 2.72 meters and weighed 6.8 tonnes.

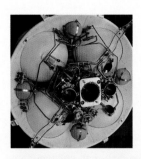

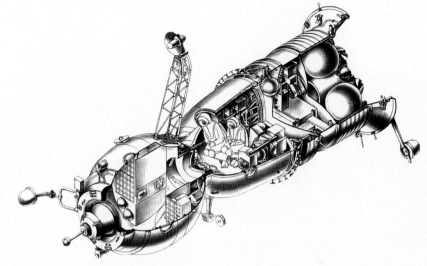

The second vehicle was the automatic Progress that could transport payloads of up to 2,300 kilograms. The lower part housed the engines. Progress was 7.94 meters long, and the pressurized part was 2.2 meters in diameter. Once its mission was over, Progress disintegrated in the atmosphere. It could also carry a Raduga capsule to return 150 kilograms of material to Earth. The capsule was 1.4 meters long with a diameter of 78 centimeters and weighed 350 kilograms unloaded.

The third vehicle, Cosmos 1443, was a habitable module 13 meters long with a diameter of 2.9 metres (50 cubic meters for the manned section). It weighed 20 tonnes. The forward section was equipped with a

RADUGA

Merkur capsule (originally habitable) for reentry. It weighed 3,800 kilograms and was 2 meters long (without the upper rescue rockets) with a diameter of 3 meters. Cosmos 1443 was launched with a Proton rocket.

SOYUZ

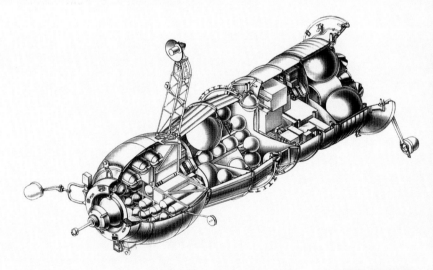

PROGRESS

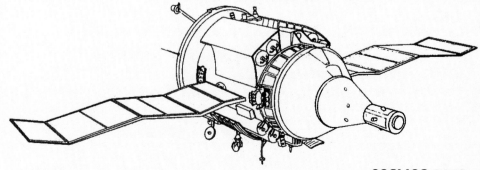

COSMOS 1443

nearly 10 years, fell out of favor and was replaced in May. His main fault, according to his enemies, was having lost the race to the Moon, and the successes of the first Salyut stations were not enough to cancel the errors of the past. Mishin certainly did not have Korolev's charisma or managerial skills or the diplomacy and determination necessary to make himself respected in the palaces of the Kremlin. But despite all that, he continued to occupy the most important post in the Soviet space program. "He was like a cat with nine lives," wrote Road Sagdeev, director of the Moscow Institute of Space Physics (IKI).

The key to his survival was his important protector on the Central Committee, Andrei Kirilenko, who, despite being in only the second rank of Party power, had his son-in-law, Yuri Semyonov, working for Mishin. And Mishin, obviously, did not hesitate in making him important, guaranteeing the future up to at least 1974, when changes in the balance of power led Marshal Dmitri Ustinov, head of the powerful military-industrial apparatus at the Ministry of Defense, to remove him.

To take Mishin's place, there arrived another historic rival of Korolev after Chelomei. This was Valentin Glushko, an expert in space propulsion and a grand egocentric with boundless ambition. His succession in May 1974 was a revenge Glushko had long sought. But to ensure his future, he kept Andrei Kirilenko happy by immediately bringing Semyonov into his circle of collaborators; and so the son-in-law continued his rise.

In the meantime, 1974 was a good year for the mil-

itary space program under Chelomei's team, as they achieved their first success with Almaz 2/Salyut 3, launched from Baikonur on June 24. Considerable improvements had been made over the design of Almaz 1. First, the solar-cell panels that supplied a total of 5 kilowatts of power were no longer fixed at the stern but rotated 180° so as to follow the Sun during orbit. In the past, the entire station had to be oriented for this purpose. The new system kept the orbital base stable, as was required for making surface observations (mainly for military reconnaissance).

The interior had been furbished differently too; the residential area was now in the narrower forward section. This was separated from the larger section by a sort of wall composed of containers for various instruments. On one side, there was a kind of corridor that united the two areas. In the bigger space was a large, cone-shaped structure housing cameras with a focal length of 10 meters whose principal function was to serve defense needs. They were also to be used, to a lesser extent, to observe the Earth's surface for scientific purposes. If pictures were needed quickly, they could be developed onboard, scanned and transmitted to Earth. On the other hand, "ordinary" film was sent to the ground in capsules with a parachute that opened 8.4 kilometers above the ground. These capsules were shipped from the lower opening of Almaz 2's docking compartment. This was in the tail of the station, and solar panels were attached to its sides. In the upper part of the compartment was a

Cutaway view of Salyut 4 interior showing the command post and the cone for scientific instruments.

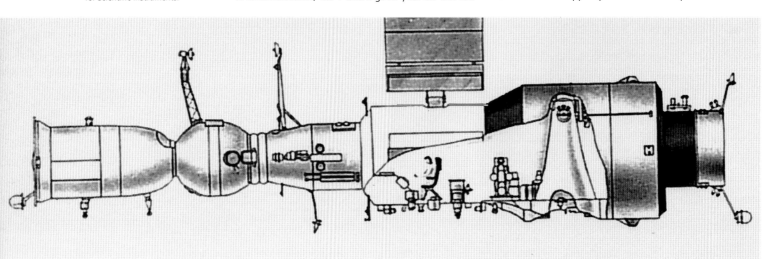

door for the cosmonauts to exit for extravehicular activities, but it was never used.

The compartment wall opposite the docking mechanism had a door into the station and faced the larger section. Here, besides the camera unit, a shower could be erected when needed. Use of the latter proved to be anything but simple.

The floor of the station was covered with Velcro to help the cosmonauts stand still in zero gravity. Two berths, one fixed and one mobile, were available for sleeping. In the forward leisure area were four portholes to render the environment less oppressive. A magnetic chess board, a small library and a cassette recorder were provided to pass the time.

The workday was interrupted by two hours of physical activity on a treadmill. To make the return to Earth less traumatic, a Pinguin suit covered the lower part of a cosmonaut's body up to the waist; decompression forced blood to circulate in the lower limbs.

To reduce the quantity of water needed to be hauled from Earth, water-regeneration experiments continued on the station using the Priboi system, which was able to condense water from the cabin atmosphere. This was an advancement in efforts to make the orbital base more autonomous.

Moreover, to save rocket fuel, the station was equipped with a system of large gyroscopes (similar to those on the American Skylab station). When these were shifted with electric motors, the flight attitude could be changed. Among other innovations were a system to improve heat regulation, equipment for electronic computing and inertial navigation and the use of Molniya telecommunications satellites as signal relays. This considerably increased the ability to communicate with the Kaliningrad control center, which before had been limited to windows of opportunity when the station and ground antennas could "see" each other.

The first crew, code-named *Berkut* (Golden Eagle), left from Baikonur on July 3, consisting of cosmonauts Pavel Popovich and Yuri Artyukhin. Their main tasks were to film special military sites and to test the new station. Their stay lasted 16 days, and reentry took place on July 19, 140 kilometers southeast of Dzhezkazgan, in Kazakhstan.

The following mission, with the Soyuz 15 capsule, was not to be so successful, however. After launch on August 26, 1974, it failed to dock.

From then on, Salyut 3 continued its orbital life in solitude, because no other crews were sent to it.

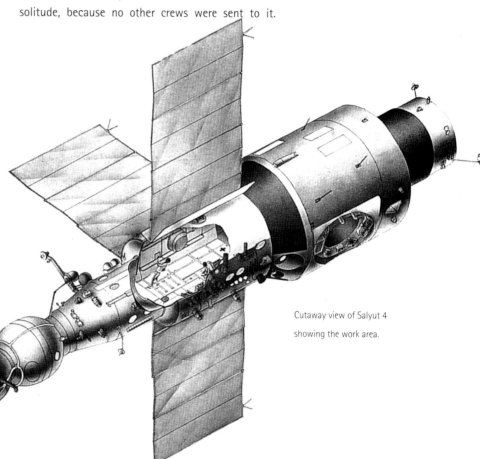

Cutaway view of Salyut 4 showing the work area.

However, it continued to be controlled from the ground, and livable conditions were maintained onboard. On September 23, the command was given to expel a capsule that was recovered on Earth. It contained the films taken during the flight and indicated the end of the station or, rather, its definitive abandonment. According to some sources, Salyut 3 should have remained active for 90 days, but it actually went way beyond that to seven months. On January 24, 1975, it at last reentered the atmosphere and broke up over the Pacific Ocean.

THE CIVILIAN SALYUT 4, A STEP FORWARD

On December 26, 1974, while Salyut 3 was still orbiting the Earth with a few weeks of life left, the new, civilian Salyut 4 station lifted off from Baikonur. Also built at the Korolev Center, its structure and dimensions were similar to Salyut 1, but its systems were almost identical to those of the station that had failed the preceding year (Cosmos 557).

Despite the rivalry between Mishin and Chelomei, however, some exchange did take place, and the new station included many improvements tested on the earlier military base. An advance had been made in power generation by installing three solar-cell panels in the narrower section of the station. Their 60-square-meter surface area was greater than that of the four panels used before and therefore supplied more power—up to 4 kilowatts. Moreover, the three panels could rotate 340° to follow the Sun constantly and make the base more autonomous. On the other hand, the system of large

gyroscopes for position control was still not perfect, and the engines had to be used more often for attitude correction, with a consequent penalty in fuel consumption.

Several refinements had been made onboard. To make management of the cosmonauts' work easier, a Stroka teletype machine had been installed, and every morning, it printed out the program for the day, prepared and transmitted by the Kaliningrad control center. Previously, the cosmonauts had had to take down their tasks by hand. A flexible rubber tube had been extended to reach the Soyuz capsule when it was docked to the station and to provide adequate ventilation.

A semiautomatic Delta system was introduced for navigation, a big improvement on the past, when the space station's trajectory had to be tracked during every orbit by a network of ground stations. Instead, Delta used a group of solar sensors and an Argon 16 computer to combine data collected by the sensors with that obtained from a radioaltimeter and to cal-

culate the orbital parameters. This allowed the station's position relative to the Earth's surface to be determined within a 3-kilometer margin of error, and altitude within a margin of a few hundred meters. The result was displayed on the Globus navigation monitor of the base's command console.

At the same time, a Kaskad attitude-control system was also introduced. This used a pair of infrared sensors, with Earth as the reference, plus an ion sensor that determined position relative to the direction of flight. In this way, the proper orientation was maintained to within 5°. Apart from increasing flight safety, the combination of the Delta and Kaskad systems simplified tracking operations from the ground and also reduced the need for doing this type of work onboard. (On Salyut 1, it had become burdensome, taking up nearly 30 percent of the available time.)

To control internal environmental conditions better, the exterior surface of the station was equipped with layers of synthetic material covered with aluminum film to reduce heat loss. Moreover, a complicated system of radiators had been adopted that gathered heat from the sunlit part and released it on the shady side. The whole thing was completed by a network of equipment, commanded by a computer, that heated or cooled the various areas of the station when they were in shade or sunlit. Finally, when the Soyuz capsule was docked, its systems were shut down and the station took care of its internal conditions.

Life onboard for the cosmonauts was still somewhat spartan, although some degree of comfort had been introduced with new technologies. A dining table with hot and cold water dispensers was located in the command area, and the cumbersome cone housing the telescopes was placed in the roomier rear compartment. Above this, on the ceiling, there was medical and exercise equipment, and in the forward area, above the guidance and control console, were sleeping bags. The toilet was on the floor at the rear of the station.

As a laboratory, Salyut 4 was abundantly equipped with scientific instruments, with astronomical and Earth observations being favored. For astronomy, there was an OST 1 solar telescope, with a 25-centimeter diameter and a focal length of 2.5 meters, housed in the infamous

cumbersome cone that rose from the floor in the larger compartment. Attached to the telescope, which was built by the Crimean Astrophysical Observatory, were a diffraction spectrometer for the first studies of the Earth's ozone layer and an infrared spectrometer for atmospheric studies. The cone also housed two instruments (an RT 4 telescope and a Filin spectrometer) for X-ray observations of the heavens. Finally, a piece of apparatus called *Silya* (Power) opened a brand-new line of research into the nature of cosmic rays.

In addition to their astronomical investigations, the cosmonauts used an Oasis space greenhouse and Biotherm incubators to conduct experiments in the cultivation of onions in a closed ecosystem and to monitor the development of frog, fish and drosophila (fruit fly) eggs. At the same time, they analyzed the behavior of their own organisms and used an exercise-cycle ergometer to keep up their muscle tone. While they pedaled the latter, they also generated electricity that was stored for future use. Nothing, in fact, was wasted.

SALYUT 4 CREWS

Salyut 4 had been in orbit for nearly three weeks when the first crew left from Baikonur to go aboard. On January 12, 1975, Alexei Gubarev and Georgi Grechko were launched in Soyuz 17, with the code name *Zenit* (Zenith), docking with the station at an unusually high orbit of 350 kilometers. The docking proceeded so normally that the crew's heartbeats were actually lower than that recorded during ground training. When the cosmonauts entered the station, they found it quite cold inside, so they turned on the heaters and the temperature soon reached 23°C. Indeed, all the crews preferred a rather overheated climate during the first days of their stay in orbit.

The scheduled activity began, with six days of work followed by one of rest. For an hour each day, the cosmonauts' lower limbs were depressurized in a Tchibis apparatus to ease blood circulation in the lower body. This was then reduced to one hour a week but later increased to a daily session during the 10 days preceding reentry. Still, even this intense treatment proved to be insufficient.

For the first time, the station had an exercise cycle,

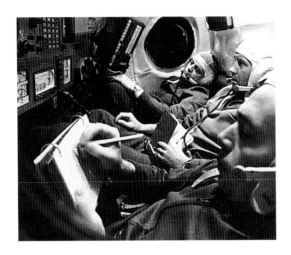

a veloergometer, the use of which turned out to be quite difficult; after a few minutes, the cosmonauts were drenched in sweat because inadequate ventilation meant that they retained body heat. Large napkins had to be used after a short time to dry off the abundant and annoying perspiration.

The principal objective of the mission was astronomical observation, but at a certain point, the crew had to deal with an emergency when the OST 1 solar telescope stopped working because of faulty alignment of its secondary mirror. They rather cleverly fixed it, but the troubles were not over; being improperly pointed toward the Sun caused a failure of the directional system. Once more, however, the crew managed to reprieve the situation so that observations could continue. After a few weeks, the ozone present in the atmosphere at the station's altitude had damaged the telescope mirror to such an extent that it was practically unusable. So together, the cosmonauts and the ground technicians invented a simple but effective method to fix it: a piece of aluminum was electrically melted in front of the mirror, and thanks to the vacuum of space, as the metal vaporized, it deposited on the mirror's surface, renewing it perfectly.

Gubarev and Grechko stayed onboard for 30 days, surpassing past records, and on February 9, they boarded the Soyuz capsule and returned to Earth, using a double maneuver made necessary by the high altitude. First, after de-docking from Salyut 4, they descended to 200 kilometers, then they began the final descent procedure by braking orbital velocity. The landing took place in late afternoon, with the crew in good physical condition and excellent humor. All the same, in the following days, both men had difficulty walking, and Grechko suffered from spasmodic pains in the chest, caused, it was discovered, by a slight movement of his heart during the absence of gravity. The specialists realized that they would have to intensify the program of physical activity in orbit to neutralize the effects of weightlessness and would also have to use some drugs.

A second crew, code-named Ural and consisting of Vasily Lazarev and Oleg Makarov, left on April 5 with Soyuz 18A. The "A" was added later, because the mission ran into almost immediate problems and never came off. Indeed, disaster was averted by a whisker when the first stage of the rocket failed to separate completely from the second. In signaling the problem, the onboard computer activated the emergency system and ignited the rockets that freed the capsule containing the crew. This took place at an altitude of 180 kilometers, with the rocket traveling at 5 kilometers a second—a high speed, but not sufficient to reach orbit. So after a suborbital flight, Soyuz 18 landed 1,600 kilometers away in the snows of the Altai Mountains of western Siberia. Lazarev and Makarov experienced deceleration forces of 15g (that is, 15 times their weight). Upon exiting, they found themselves in dangerous circumstances, since they had no way of keeping warm, although they did have food for a few days. While they waited, they gathered wood and started a fire. Luckily, their descent had been tracked by radar, and rescue teams appeared on the horizon within a few hours to take them to safety.

On May 24, Piotr Klimuk and Vitaly Sevastyanov boarded Soyuz 18B (again numbered "18" after the failure of 18A) under the code name Kavkaz, and left from Baikonur to dock with Salyut 4—this time with no surprises. The station had by now been in orbit for five months and was in need of some maintenance, including replacement of the air filters, repair of faulty instrumentation and the changing of six condensers. These operations proved to be easy and confirmed the validity of the specifications used for the installations inside Salyut 4. The maintenance protocol had been planned in detail, using methods that gave the cosmonauts easy access.

After that, celestial and terrestrial observations continued with 90 scientific and technological experiments. They included observations of the Sun during a period of intense solar activity. However, these could not be completed because of an unfavorable alignment of the station's orbit.

Despite all the adjustments, Salyut 4 was no longer in optimum condition for habitation: the environmental control system was not working perfectly any more, and mold was spreading on the walls. So the two cosmonauts suggested cutting back the mission. But at Kaliningrad, the mood was for continuation to

Above, Nikolai Rukavishnikov (left) with Georgi Ivanov of Bulgaria.
Left, Valery Ryumin (left) with a companion during a ground simulation.
Below, three cosmonauts undergoing training in the Salyut station simulator at Moscow's Star City.

show the world what the Soviet space program was capable of.

Thus, on July 15, while the Soyuz 18B cosmonauts were still aboard Salyut 4, the Soyuz 19 capsule left Baikonur with Aleksei Leonov (the first space "passenger") and Valery Kubasov aboard. Two days later, a historic event took place, as Soyuz 19 docked with the American Apollo 18 capsule carrying Thomas Stafford, Deke Slayton and Vance Brand. This was the famous Apollo-Soyuz joint mission carried out by the United States and the Soviet Union as the symbol of a union that would in fact take much longer to become reality. On July 17, after the docking between the two capsules, the united crews talked via radio with cosmonauts Klimuk and Sevastyanov in Salyut 4. At this point, the USSR had two crews in orbit: one on the international flight (tracked from Kaliningrad) and one in Salyut 4 (tracked from Yevpatoriya in the Crimea). It was indeed an exploit to show the world.

On July 26, the crew of Soyuz 18B was authorized to

return after a record 63 days in space. Two days later, Soyuz' engines were used to raise the station's orbit to 350 to 370 kilometers for the first time so that the flight could be continued by remote control from Earth.

But the station was not finished yet. On November 17, 1975, an unmanned Soyuz 20 capsule was launched from Baikonur that, after a long pursuit, docked automatically with Salyut 4 during the 34th orbit instead of the usual 18th. The object of the mission was to put the Soyuz systems through an endurance test.

Salyut 4 was judged to have performed extremely well from the point of view of both science and technology. De facto, it was the first station able to sustain prolonged crew activity as well as to experiment with a series of new features that would lead to the design of a more complex second-generation station. Until then, the main aim had really been just to learn how to build an orbital base and to invent the brand-new equipment needed to do so. Now, basic station engineering was available to take the next step forward, to build a base that could make routine missions in safe conditions possible. This was to happen with Salyut 6.

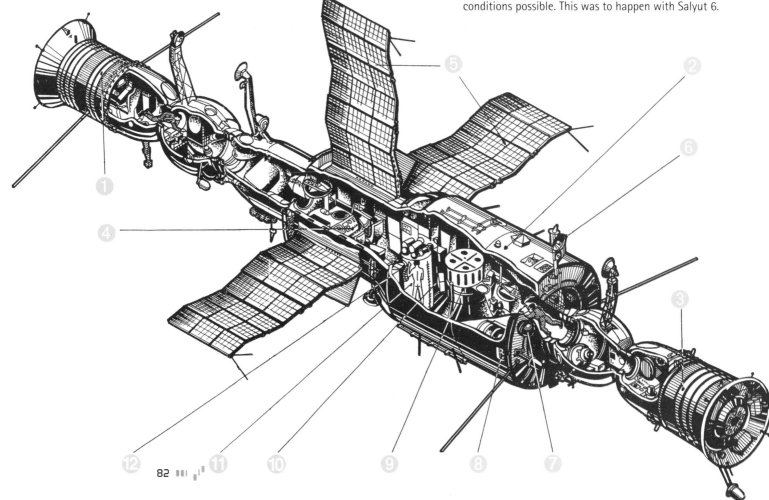

Salyut 6 in orbit.

CHANGES IN SOVIET SPACE POWER

The year 1976 was an important year for the Soviet space program with regard to things taking place on the ground. In the early months of that year, Marshal Andrei A. Grechko suffered a fatal heart attack. For almost 10 years, he had been defense minister and the major supporter of Vladimir Chelomei as head of military space programs. For this, he was known as "the costliest man in the Soviet Union." Now the turn of events was against Chelomei, and it was the beginning of the inexorable end.

After Grechko's death, his place as defense minister was taken by Marshal Dmitri Ustinov, hitherto head of the military-industrial commission, who had put Glushko in Mishin's place as director of the Korolev Center two years previously. A few weeks later, during the 25th Soviet Communist Party Congress, Ustinov, backed by Kirilenko, secretary of the Central Committee, promoted Glushko to membership in the Party Central Committee, in the process making him the most power-

ful man in the Soviet space program. Indeed, besides having control of almost all the enormous space empire in his hands, he also had political power that he used immediately, obviously to his own advantage.

Without losing any time, Glushko decreed the end of the Almaz military space-station program and shut down Chelomei's project. Immediately afterwards, Chelomei had to watch helplessly as the new defense minister annulled all the space contracts for which he was responsible. To humiliate him further, Ustinov even ordered the destruction of expensive elements of an orbital radar station that Chelomei had been building. Chelomei was left with nothing, and from that moment on, he was limited to working with a few assistants on research on military space defense systems that were very similar to the famous "space shield," proposed later by the American General Daniel Graham, which became the basis for the "Star Wars" plan initiated by President Ronald Reagan.

So Glushko saw his last adversary destroyed, and to

The simulators at Star City, near Moscow, for cosmonaut training: the Soyuz capsule simulator (foreground); the Salyut station (background).

wipe the slate of history clean, he took another symbolic and most significant step: he officially canceled the N-1 lunar rocket program—which had, in any case, been practically abandoned after its fourth disastrous test in December 1972.

SALYUT 5/ALMAZ 3, THE LAST MILITARY STATION
While Chelomei exited the stage, construction of the last military station, Salyut 5/Almaz 3, was being completed. It was launched on June 22, 1976, from Baikonur into the traditional low military orbit of 275 kilometers. Almaz 3 was practically the same as Almaz 2, except that—according to reliable Western sources—a spherical transfer module was placed at the front, equipped with four docking systems. In practice, this was the first test of a multiple docking module similar to that later used on the Mir station. Another difference was inside, where the walls and the rough edges of the equipment had soft covers to make the cosmonauts' movements less risky, as the cosmonauts themselves had suggested.

The first crew, code-named Baikal, took off for the station on July 7 with Soyuz 21. They mainly conducted photographic reconnaissance of the surface below, using a camera similar to the previous ones but with some improvements. The cosmonauts, Boris Volynov (on his second flight) and Vitaly Zholobov, were both officers, and in addition to their military tasks, they also carried out secondary geological observations, scientific experiments on biological organisms and technological experiments with welding and soldering in zero gravity.

However, on August 24, 1976, the news agency Tass announced surprisingly that the crew was returning to Earth after 49 days, 10 days earlier than expected, and without carrying out the traditional exercises for readaptation to gravity. Naturally, no precise information was forthcoming, but the hasty return was confirmed by the fact that it took place at night, 250 kilometers away from the usual landing area in Kazakhstan, on the land of the Karl Marx collective farm.

The official bulletin announced that the crew was in "satisfactory condition," a declaration that was rather more understated than usual and thereby raised the suspicion that the hasty departure from orbit was due to the cosmonauts' physical condition. In addition, on August 18, the newspaper Izvestia reported that the two men were suffering from generalized balance problems and that psychologists had suggested playing music to raise their spirits.

Other possible reasons put forward for the early return were a fire onboard, difficulties in internal-atmosphere control and health problems caused by fumes from the chemicals used to develop the reconnaissance film. At this point, after the abundance of rumors and theories, no one knew whether the station was about to be abandoned.

The Soyuz 23 mission departed on October 14 carrying Vyacheslav Zudov and Valery Rozhdestvensky, under the code name Radon. Unfortunately, Soyuz' automatic system to guide the rendezvous with the station did not work, and the flight was a failure. The landing took place the next day and was somewhat of an adventure, as the capsule not only descended during the night and in the middle of a snowstorm but also finished up in the frozen Lake Tengiz, 8 kilometers from shore. The ice-covered surface meant that the available ships could not reach the capsule, and helicopters were also unable to intervene immediately because of thick fog conditions. The capsule's batteries were dead following the flight operations and therefore unable to heat the cabin, so the two cosmonauts remained in icy cold for several hours until morning, when a helicopter managed to drag the vehicle to shore and rescue the two half-frozen unfortunates.

The Almaz 3 station had been in orbit for seven months and was therefore in an almost critical condition when a last expedition was attempted on February 7, 1977. The protagonist was Soyuz 24, carrying a veteran and a new recruit—cosmonauts Viktor Gorbatko and Yuri Glazkov (crew code-named Terek). This time, docking took place without a hitch, even though it was executed in the dark with the aid of a spotlight that lit up the target area. But the two crew members did not enter immediately as was usual, instead spending the night in their capsule. Upon waking, they put on gas masks and went into the station to check for toxic substances in the atmosphere. This confirmed the hypothesis that an accident had caused the early return of Soyuz 21.

However, the inspection showed that the environment was "comfortable," and the cosmonauts announced that it was "easy to breathe." Alexei Yeliseyev would declare later that the internal environmental control system had continued working, purifying the air prior to the arrival of the new crew.

This operation was followed by the usual military reconnaissance and some scientific experiments. Work was interrupted abruptly on February 17, because the cosmonauts had heavy colds that could become a serious danger in space. The doctors made the prosaic suggestion that both men should sit in the sunlight at a station porthole for 15 minutes. This natural cure worked.

Four days later, a direct television broadcast from space demonstrated an interesting experiment relevant to the future Progress automatic freight carriers. The cosmonauts evacuated the air from the station, replacing it with new air from the Soyuz tanks. All this was done automatically, and Gorbatko and Glazkov recounted that they felt a pleasant "light breeze" during the

exchange of 100 kilograms of air. Once the test was finished, the base was prepared for automatic operations and Soyuz was loaded with the scientific and reconnaissance materials accumulated. After only 18 days, the crew returned in the Soyuz capsule on February 25, and the next day, the automatic capsule loaded with film also de-docked and was recovered on the ground.

After a few weeks, a shift to a more circular orbit hinted misleadingly that a new crew (Anatoli Berezovoi and Mikhail Lisun) would be on its way to the station. Although the two cosmonauts were in readi-

Left, the tank at Star City for EVA training for the Salyut station.
Below, cosmonauts Romanenko and Grechko inside the simulator.

ness, their flight was canceled. The end of Almaz 3 was near. On August 8, 1977, the control center fired the station's engines to slow its motion. As a natural consequence, the station began to descend, and its journey ended with disintegration in the atmosphere above the Pacific Ocean.

And so, the chapter of the Soviet Almaz military stations closed, and a new phase of civilian operations was begun with a more evolved generation of stations that would benefit from the knowledge and experience accrued over seven years.

SALYUT 6, THE SECOND-GENERATION STATION

The year 1976 ended with two important political events. In China, Chairman Mao Zedong died in September and was succeeded by his prime minister, Hua Kuo-feng. In November in the United States, Jimmy Carter, a Democrat, was elected president, beginning a period of White House weakness on all fronts, especially in the space program. In the meantime, the first data transmitted by NASA's two highly complex robotic probes, Viking 1 and Viking 2, which had landed on Mars to look for signs of life, failed to live up to expectations on that score.

American space-program activities continued to be disappointing. NASA was increasingly held back by the difficulties of building a reusable shuttle, the first launch of which was victim to a string of delays. On the other hand, two interplanetary missions that would make space-exploration history began in

August and September with no problems. Voyager 1 and Voyager 2 started their journeys to Jupiter and Saturn. (Voyager 2 would eventually manage to reach the boundaries of the solar system after flying by Uranus and Neptune for the first time.)

September 29, 1977, was an important date for the USSR, because a Proton rocket blasted off from Baikonur carrying the first second-generation space station, Salyut 6. This marked a profound change in the history of manned orbital bases, not just because of the decidedly innovative characteristics derived from systems progressively tried and tested on preceding stations but also because of the means of transport linked to it and created explicitly for it.

Physically, the new station had the same dimensions as the Salyuts that came before: 15.8 meters long with a maximum diameter of 4.15 meters. It was still divided into two cylindrical modules, with the forward one having a smaller diameter and housing the control and guidance system. In front of this was a compartment with the docking system and its antennas.

Second docking system: The first and most important new aspect was the second docking system attached at the rear. Both the manned Soyuz capsules and the new automatic Progress vehicles under construction could berth with it. Furthermore, Progress could resupply fuel as well as food and materials, thereby achieving the objective for which the new-generation Salyut had been built: long-duration missions.

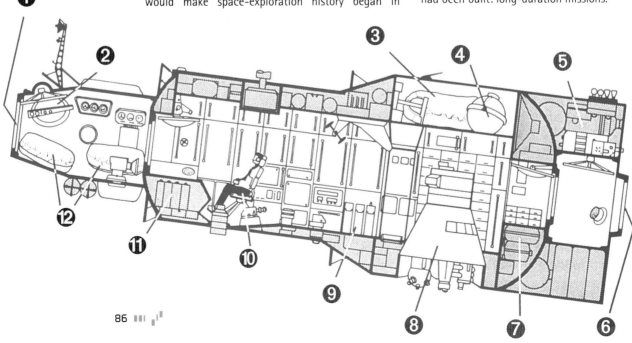

The new ODU engine system: The second major new element concerned the station's propulsion system, the rocket-engine block that allowed the station to change its orbit and correct its position. The rear docking system and the propulsion equipment were interconnected, and the solution adopted was the one tested on the Almaz military stations—a docking port in the center of the base's longitudinal axis and the two more powerful engines and their tanks at the sides: three tanks for fuel and three for oxidant. The whole assembly was integrated into a single, unpressurized, 1.9-meter-long module placed at the rear, in the middle of which was a pressurized tunnel for men and materials. The docking ring for an external vehicle also carried connections for supplying fuel and air as well as radio and electrical links.

The whole Integrated Propulsion System (called by the acronym *ODU* in Russian), including the two more powerful engines having 300 kilograms of thrust apiece and 32 small rockets for attitude control, each providing a thrust of 14 kilograms, was radically revised, sim-plified and more powerful compared with its military predecessor. First, everything used a single fuel mix to reduce the number and type of conduits. Moreover, the mix did not need an igniter, as the two propellants used would burn spontaneously when they came into contact. In addition, the turbines used to load fuel into the tanks were eliminated; in their place was a decidedly simpler and safer pressurization system using high-pressure nitrogen to inflate the elastic bladders.

The SOUD guidance and control system: The third system that had changed significantly was the guidance and control instrumentation that made up the Orientation and Motion Control System (*SOUD* in Russian). It consisted of gyroscopes, ion sensors, solar and stellar sensors, a manual sextant as well as the Delta automatic navigation system, Kaskad orientation systems and (as explained in a technical note) "radio equipment for the rendezvous, which functioned with the radio equipment of the capsule to be docked with, to measure the relevant parameters for movement."

Above left, cosmonaut Alexander S. Ivanchenkov (lower right) with two companions on Salyut 7.

This was the new Igla system that was a considerable help in approach maneuvers. All the various parts of the delicate and complicated SOUD system were backed up with reserve equipment that would come into operation in the event of a breakdown.

Solar panels: Power was generated by three large solar-cell panels installed around the narrower section of the station. The panels supplied 4 to 5 kilowatts, and each was equipped with solar sensors and motors for automatic orientation (to keep the cell surfaces pointing at the Sun). The communications antennas were placed at the ends of the solar panels so that both the power generated and the radio signals passed through a rotating joint at the base of the panel. This was also a new aspect—only the automatic orientation system had previously been tested.

Two fixed telescopes: The various instruments on the station (some of which could be switched during missions according to the type of research to be done) included two fixed telescopes. These were located in the cumbersome unpressurized cone in the wider work area of Salyut 6, taking up a lot of space. The first telescope (BST-1M) had a fairly large, 1.5-meter-diameter mirror and was dedicated to observing the sky in infrared, ultraviolet and submillimeter ranges. Although it was exposed to the vacuum of space, the telescope was cooled further by a cryogenic system that kept it at −269°C to avoid possible disturbances caused by ambient heat sources. In any case, its use was limited to the shaded periods of orbit; the rest of the time, it was protected by a cover. The second astronomical instrument was a Yelena telescope for the study of gamma-ray sources.

Two Earth cameras: A multispectrum MKF-6M camera with a Zeiss lens, previously tested on Soyuz 22, was onboard to study the surface of Earth. At the customary flight altitude of 355 kilometers, an area of 165 by 220 kilometers could be photographed with a resolution of 20 meters. The camera could take six pictures simultaneously in as many different spectral ranges: four in visible light and two in infrared. The

film was prone to deteriorating rapidly in space because of the intense radiation, so the crews visiting the station regularly brought the film back to Earth.

The second instrument (KATE-140) was a camera to make stereoscopic topographical maps in both visible and infrared. Each photograph contained an image of 450 square kilometers with a resolution of 50 meters.

Salyut 6 also had equipment for biomedical research concerning humans and other organisms brought onboard, as well as experiments in plant growth and materials science (using Splav and Kristall furnaces). Twenty portholes along the walls of the station allowed the cosmonauts to admire the view and, above all, to work with the different cameras.

Life onboard: To make life onboard comfortable, the designers had given much attention to reducing vibration and noise generated by the various systems. This followed a suggestion from the cosmonauts, who had explained that although noise helped them notice when something was not working properly, it became irksome over the long term.

Two berths were provided for sleeping, one on the ceiling and one on a wall. For hygiene, the usual plastic-curtained shower that could be raised when needed was placed in the rear. The water used for washing was repurified, mineralized and then used to reconstitute the cosmonauts' dehydrated food.

The cosmonauts' diet provided 3,200 calories a day, with a menu that could be changed daily. About 65 types of food had been prepared and stored onboard, although supplies were renewed with the arrival of the Progress automatic freighters. To stave off dehydration, the cosmonauts had to drink two liters of water a day. Books, videos and a magnetic chess board were available for relaxation periods.

The designers had planned an orbital life for Salyut 6 that ranged from 18 to 24 months to allow at least three missions of increasing duration (90, 120 and 175 days). Brief visits of a few days' duration were scheduled to make the necessary exchange of Soyuz capsules. The capsules' stay in orbit could not exceed a three-month safety margin. In making these provisions, the Korolev Center, directed by Glushko, chose to be cautious, as the technicians were aware that a series of systems was being introduced for the first time for a prolonged, previously untested period of use. Thus, it was claimed that if even 50 percent of the objectives were met, the program could be considered a success. As it turned out, despite initial difficulties, the performance was better.

FIRST MISSION: SUSPENSE AND FAILURE

On October 9, 1977, Soyuz 25 left Baikonur with the first Salyut 6 crew (code-named Photon), consisting of Vladimir Kovalyenok and Valery Ryumin. Soyuz 25 docked with the central unit, but the attachment mechanism did not activate, preventing a solid docking. Nothing could be done except to de-dock and immediately return to Earth after 48 hours and 46 minutes, because the onboard batteries did not permit long stops or further attempts.

What had obstructed the union between the two vehicles? Was there a problem in the station's docking system or the capsule's? The latter could not be checked, because the forward part was jettisoned before reentry. All that could be done was to check the station; and so, on December 10, 1977, a new

crew—Yuri Romanenko and Georgi Grechko—departed with Soyuz 26. Grechko was not only a specialist in docking systems but also a space-station expert, having stayed on Salyut 4 for a month.

This time, docking took place at the rear section and without difficulty. Once onboard, the two cosmonauts made a space walk to inspect the conditions of the forward docking collar. "Everything is in order," Grechko said after 20 minutes of inspection that he later declared "very difficult" but "extremely satisfying." The problem, therefore, was with the Soyuz, and the Kaliningrad control center breathed a collective sigh of relief. A similar problem with the station would have seriously compromised its future. Vladimir Shatalov, as chief of the cosmonaut corps, decreed that from this point on, preparing for extravehicular work (using a large swimming pool to simulate space conditions) should be an integral part of ground training.

Cutaway drawing of Salyut 7:
1) Soyuz capsule
2) connecting compartment
3) work compartment
4) container for scientific instruments
5) docking systems
6) connecting passage
7) Soyuz capsule
8) systems compartment
9) shower
10) control panel

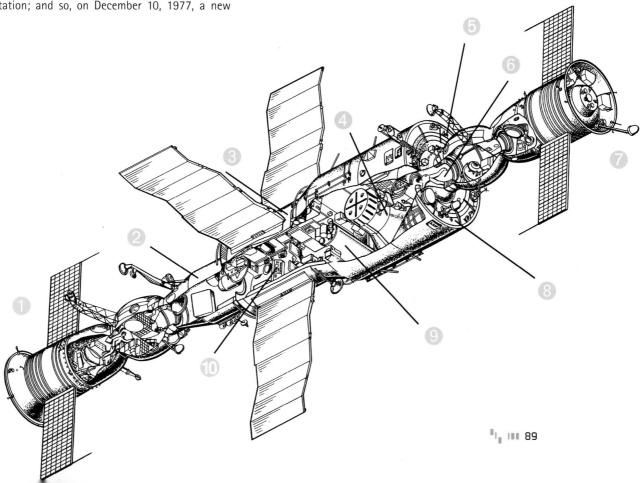

Top, forward docking system
of Salyut 7.
Above, protective cone used
during launch of Salyut 7.
Right, Salyut 7 during ground
fitting.

Soyuz 27 to test the forward docking mechanism. As a precaution, Grechko and Romanenko put on their space suits and shut themselves inside their Soyuz 26 capsule, ready for departure in case of danger. In particular, it was feared that the impact of docking at 30 centimeters a second (generating a brief 40g shock) would negatively affect the rear docking system. Nothing untoward happened, however, and the two crews soon found themselves together inside Salyut 6. It was the first time that two manned capsules had been joined to a station at the same time. Another step forward had been taken, and now, a cosmic convoy 30 meters long and weighing 33 tonnes had been formed in orbit.

One of the scheduled experiments, named *Rezonans*, concerned the damage that structural resonance might cause in uniting different space-station components. To check this out, the Kaliningrad control center ordered the cosmonauts to jump up and down in various parts of the station while instruments recorded and transmitted to Earth the vibrations and the reactions of the cosmic convoy. It proved to be extremely solid, so resonance would not be a problem: the station could grow.

After five days, on January 16, the new arrivals left on the old Soyuz 26, leaving the new Soyuz 27 with the station. From then on, this exchange of capsules would become the norm. Once back on Earth, the crew had to apologize to their colleagues still in orbit for having forgotten to take home the letters they had written to friends and relations.

FIVE LONG-DURATION MISSIONS

Between December 1977 and May 1981, 33 cosmonauts stayed in Salyut 6. Ten of these participated in long-duration missions of 75 to 185 days. Eleven crews paid short visits of 4 to 13 days. And then there were the visits by foreign guests.

The success of Grechko and Romanenko, who concluded their mission with a landing on the snows of Tselinograd on March 16, 1978, after having been visited by two crews and the first space-station resupply by the automatic vehicle Progress 1, led to an extension of the sojourns. Further, there had been no nasty surprises concerning that most important question—

The two cosmonauts began a long stay of 96 days, establishing a new record. Their day was synchronized with Moscow time, beginning with a wake-up call at 8 a.m. and ending with bedtime at 11 p.m. During rest periods, Grechko, having a livelier temperament, retired to the forward docking compartment to draw and to take photographs, while the more docile Romanenko preferred to stay at the station's control panel.

They toasted the New Year with fruit juice, and on January 11, 1978, the first visiting crew of Oleg Makarov and Vladimir Dzhanibekov was on its way in the reaction of the human body to long stays in space.

When Grechko and Romanenko landed on Earth after three months, their condition was "decent," according to the medical examiners. However, they did find it difficult to walk or even lift a cup of coffee, and their minds were still in orbit. Upon awakening on the second day, they tried to get out of bed by floating as they had done on the station in zero gravity. The doctors also discovered that their hearts had diminished in volume because they had not had to work so hard. The number of red corpuscles in their blood had also decreased. But by the fourth day, the physical conditions of both men had returned to normal.

THE FIRST GUEST:
A CZECHOSLOVAKIAN COSMONAUT

Among other things, Grechko and Romanenko's mission had inaugurated a new "political" tradition for Soviet space missions: that of sending citizens from "friendly nations" into orbit. The first such visitor was Vladimir Remek, a Czechoslovakian researcher who took off in Soyuz 28.

Then on June 15, 1978, a pair of cosmonauts was launched in Soyuz 29 who would extend the record stay in orbit even further. The crew (code-named Photon) consisted of Vladimir Kovalyenok, who had been part of the first unlucky flight to Salyut 6 with Soyuz 25, and Alexander Ivanchenkov. They adapted to weightlessness more easily than their predecessors had, and after only three or four days, their condition was considered to be normal, and the usual daily routine began: physical exercise, scientific experiments, biomedical tests and observations of the sky and Earth. The Progress 2 vehicle that arrived on July 9 brought not only supplies, materials and spare parts but also a cosmic-ray experiment that had to be

Interior of Salyut 7, with the command post in the foreground.

Onion grown aboard Salyut 6.

installed on the exterior of the station. Twenty days after their arrival, the cosmonauts went outside: Ivanchenkov worked, while Kovalyenok filmed him with a color TV camera.

ENCOUNTER WITH A METEORITE DURING A SPACE WALK

When the station was in shadow, the two cosmonauts stopped and rested for about 40 minutes (an acceptable pause, given the fatiguing nature of extravehicular activity), during which they could admire the black sky and the fixed, unflickering lights of the stars as they had never seen them from Earth. At a certain point, a flash of light and his companion's shout in his earphones made Ivanchenkov turn his head, and he could see his friend's shadow moving excitedly. Then they realized that a meteorite had crossed Salyut's path. With the return of daylight, they finished the job and were invited to go back inside. "It would be nicer to stay here," joked Kovalyenok. "After all, it's the first time we've been out of the house in 14 days." At the control center, this message was understood and accepted. The space walk concluded after 125 minutes, and the doctors commented that it had acted as a tonic for the crew.

POLISH AND EAST GERMAN VISITORS

The mission then saw the successive arrival of two foreign guests, Miroslaw Hermaszewski of Poland and Sigmund Jahn of East Germany, and three Progress resupplies. First, however, the aft docking port had to be cleared of Soyuz 31. Therefore, Kovalyenok and Ivanchenkov entered the capsule and moved it 200 meters away from the station. Ground control turned Salyut 180° to allow Soyuz 31 to dock at the forward end, leaving space at the rear for the Progress freighter. The two cosmonauts concluded their mission on November 2, 1978, using Soyuz 31 to return home after 139.5 days and landing softly 180 kilometers from Dzhezkazgan, in Kazakhstan. "When I stepped out of the capsule, I smelled the ground warmed by the Sun. I will remember it forever," Ivanchenkov said.

Their record stay in space had reached four months.

At the beginning of their stay, the cosmonauts had suffered from insomnia and headaches caused by an excess of carbon dioxide (this taught them to change the filters more often). They had lost weight, and their bones had decalcified—all anomalies that would prove to be the norm in space flights of a certain duration. All the same, recovery on the ground was quite quick: the day after landing, the two cosmonauts were swimming in a pool, and on November 4, they went for a 40-minute walk.

FUEL LEAK: FIRST REPAIRS

At this point, it was the turn of Vladimir Lyakhov and Valery Ryumin, who took off on February 25, 1979, in Soyuz 32, to break the record and also to deal with the first serious trouble and thus the first risky orbital repair. The station had been in orbit for 18 months, and some of its components were reaching the end of their technical life span. In short, the base needed a general checkup and some maintenance.

Above all, it was necessary to solve the problem of a potentially serious fuel leak that had been discovered by the controllers at Kaliningrad before the arrival of the new crew. However, they had to wait for the arrival of Progress 5, because the idea was to empty the leaking tank, for which the automatic vehicle might be needed. They first proceeded with general maintenance work before dealing with the leak by transferring part of the remaining fuel to an adjacent tank. Once this was filled, the rest was loaded into a Progress tank. In the meantime, the station was rotated so that centrifugal force aided the separation of the nitrogen from the fuel inside the leaking tank. When the operation was finished, the rotation was stopped by firing Soyuz' engines, and the faulty tank's valve was left open for a week to make sure nothing remained inside. Then it was closed, filled with nitrogen and isolated from the other fuel reserves, successfully solving a dangerous problem.

The Progress vehicle's engines then raised Salyut 6 to the usual 350-kilometer-high orbit, and on April 13, the automatic vehicle de-docked and disintegrated in the atmosphere. Work could return to its normal rhythm.

A NASTY INCIDENT FOR A BULGARIAN GUEST

A brief mission departed from Baikonur on April 10, 1979. Onboard Soyuz 33 were Nikolai Rukavishnikov and Georgi Ivanov, a Bulgarian cosmonaut (crew code-named Saturn). Everything went well until the capsule was 1,000 meters away from the station, where it should have fired its engines for six seconds to come closer. But after three seconds, the rockets automatically cut off, causing a violent shock of reaction in the capsule and its crew. Ryumin was watching the scene through a station porthole when he saw a flame shoot out from Soyuz' tail. He realized something was wrong, perhaps seriously. It was difficult to explain the cause of the accident at once, but certainly the main engine had failed. Trying to re-ignite it could cause an explosion. So, Rukavishnikov saw his chance to enter a space station slip away for the second time—on his first mission to Salyut 1, docking had failed, and he had had to return to Earth. Perhaps because of this, he suggested carrying on and attempting to dock using the small attitude-control rockets. His proposal was turned down, because it would have further complicated the situation rather than solve it. It was decided that it was better to return at once, as soon as the orbit allowed, using the reserve engines for the first time.

Rukavishnikov, worried about the risks, remained awake. Ivanov, the Bulgarian, who had already shown great poise at takeoff (his pulse had registered a neutral 74 beats a minute, as if he had been sitting in an armchair reading the paper), lowered his mustached face and went to sleep.

The next day, fortune helped them as the engine fired normally with full power. However, since it could not be regulated like the main one, the descent was somewhat violent, and the crew experienced acceleration forces of up to 10g instead of the usual 3 or 4. Landing took place in the dark, 15 kilometers away from the target area, and when the rescue teams arrived, the two cosmonauts were already out of the capsule. Only 47 hours had elapsed since their launch. His face taut with stress, Rukavishnikov told his rescuers, "I feel as if I've spent a month in space."

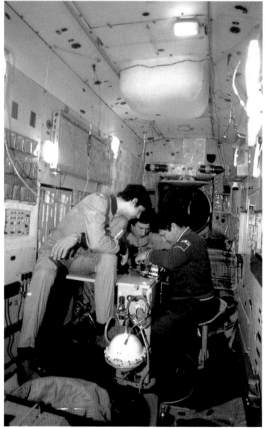

Above, cosmonaut Valentin Lebedev during a Salyut 7 space walk.
Left, cosmonauts during ground training with the Salyut station simulator.

While the roster of missions and visitors (a Hungarian cosmonaut was planned) was necessarily rescheduled, activities on the station continued without a hitch. The arrival of Progress 6 on May 15 was followed by Progress 7 on June 30, bringing supplies and some instruments as well as the first large radio telescope to be used in space.

THE RADIO TELESCOPE: A SPACE RESCUE

The antenna of the radio telescope (named KRT-10) had a deployed diameter of 10 meters, but its size when packed for transport was only 1.5 meters. The whole thing was attached to Salyut's rear docking system. A ground command de-docked the automatic Progress capsule and parked it a few hundred meters away, leaving the still closed telescope antenna free. A command from the station initiated deployment, and the parabolic dish opened, monitored by the Kaliningrad center via a color TV camera installed on the nearby Progress. After calibration, the antenna, similar in size to many ground-based radio telescopes, carried out numerous observations between July 24 and August 9 in conjunction with a 70-meter-diameter radio telescope in the Crimea. Its targets ranged from the heart of our galaxy to Italy's Mount Etna.

When the order was given to de-dock on August 9, however, the parabolic dish inexplicably remained attached to the station. This was a serious problem, since it was located at Salyut's rear docking system, preventing the arrival of a Progress resupply vehicle. "You have two possibilities," flight director Alexei Yeliseyev told the pair in space by radio. "Return to Earth to celebrate Valery's birthday [Valery Ryumin was one of the cosmonauts in orbit], or stay up there and try to free the station of the antenna. It's up to you." Both men gave the rather obvious reply and began preparing for the riskiest extravehicular operation ever attempted. On August 15, Ryumin and Lyakhov went out, attached by a safety cable and the umbilical cord that supplied electricity for the radio link. The length of the cord—20 meters—proved its usefulness for the first time.

They had to use shears to cut the telescope dish's

Top, cosmonauts aboard
Salyut 7.
Above, Salyut's hygiene
system.

A month of investigations revealed that the combustion chamber had not reached the proper pressure, and before the fuel mix could reach levels that could cause an explosion, the chamber sensor had turned everything off. There had never been problems with Soyuz' engines before, and the accident came as a surprise, leading to a review of production and the introduction of corrective measures. Obviously, no intervention could be made to prevent the same problem occurring with the Soyuz 32, already docked with Salyut 6. Moreover, its period of exposure to the space environment was approaching the safe limit of three months, and the cosmonauts could no longer use it.

This meant that a new capsule, Soyuz 34, had to be sent up on an automatic flight. It took off on June 6, 1979, and Soyuz 32, full of specimens, film and instruments, returned to Earth on June 13. (But first, the cosmonauts' individually made-to-measure seats were transferred from Soyuz 32 to the new vehicle attached to the station.) Although its safe-duration limit had been slightly exceeded, reentry of Soyuz 32 was perfect.

net at four points, then they pushed it away with a pole. The station was in shadow when they went outside, so they prepared to make the cut as soon as the Sun rose to light the scene. But when the first rays hit Ryumin's helmet, he had to lower his visor, thereby lowering his helmet's internal temperature and misting up his vision. Clearly not ideal conditions to work safely and well. All the same, work proceeded until Ryumin informed the Kaliningrad center at last that "the antenna's gone." He heard the controllers' applause in his headset. After 83 minutes, the cosmonauts reentered the station satisfied that they had managed to save Salyut from the specter of being abandoned.

On August 16, they celebrated a birthday in orbit and prepared for their return home by activating the systems for automatic flight. On August 19, 1979, they entered Soyuz 34 and landed just after noon as planned, 170 kilometers from Dzhezkazgan.

Their flight had lasted 175 days without physical or psychological problems. According to the doctors, this was due to having increased the time for physical activity by 15 percent. Thirty minutes after landing,

they felt in better condition than their predecessors. However, they were pale, dizzy and had some difficulty with balance and talking. The acute phase of readaptation lasted three days, after which their condition improved, thanks to the new techniques (pool exercises, massages, saunas and specific drugs) that accelerated the reacquisition of normal functions. Tests showed Ryumin to be exceptional—he was the only cosmonaut to put on weight (700 grams) instead of losing it as everyone usually did. His rapid return to space was probably helped by this peculiarity.

The months went by, and the station orbited unmanned. But it did receive a silent visit when, on December 16, after three test missions under the generic name Cosmos, a new, more evolved and powerful version of the Soyuz capsule, capable of carrying three cosmonauts, was tried and proven by an unmanned, automatic docking with Salyut. This was Soyuz T-1 ("T" for "Troika," the third-generation Soyuz), which remained attached to the orbiting base for more than three months, until March 23, 1980. A week later, Progress 8 delivered supplies, hinting at a new long-duration mission.

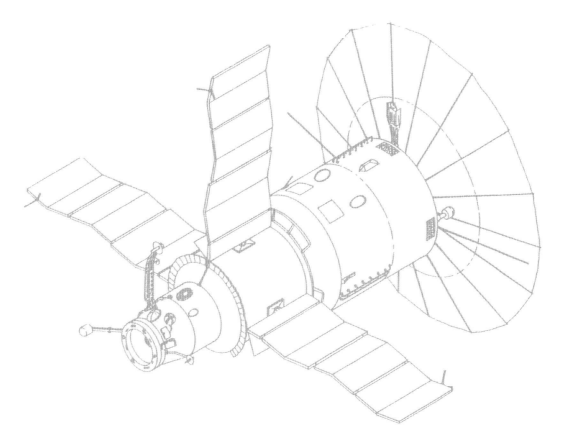

Sketch of Salyut 6 with the KRT-10 radio telescope mounted at the rear.

Cosmonauts Svetlana Savit-skaya (front), Vladimir Dzhanibekov and Igor Volk in Salyut 7 in 1984.

FOURTH RECORD MISSION

On April 9, Soyuz 35 departed with the "Dnieper" crew, consisting of Leonid Popov and, again, Valery Ryumin. This choice surprised and disappointed the reserve pair, as Popov should have been accompanied by Valentin Lebedev. Lebedev, however, had been injured during training and had needed surgery. Ryumin was preferred over the reserve crew; he was asked to return to space only seven weeks after his last record stay. Ryumin was willing but expressed one reservation: "How will I tell my wife?" In the end, he went.

The first week was spent on maintenance and repairs, while Popov got over the usual troubles of adapting to space. Then they turned to the customary research activities, and their talk was enlivened by tales of a cucumber that they found to have grown on the station, as well as orchids, chosen because they could flourish in a dry atmosphere like the base's. However, those that were brought already blooming from Earth wilted, while those sown onboard grew abundantly and with such strong roots that they burst out of their container.

Another Progress vehicle brought supplies on April 29, and on May 1, the cosmonauts were ordered to take a break to watch the live TV broadcast of the parade in Red Square and to talk to friends and relatives at the Kaliningrad control center, also via live TV. At the end of the month (May 26 to be exact), it was the turn of another guest, Hungary's Bertalan Farkas,

who arrived on Soyuz 36. For eight days, he and his crewmate, Valery Kubasov, were involved mainly in conducting medical and pharmaceutical (interferon production) experiments. Then they returned to Earth.

June 6 saw the turn of the crew of Soyuz T-2, Yuri Malyshev and Vladimir Aksyonov. They wore lighter, more elastic, less cumbersome space suits. The space saved thereby meant that a third passenger could be reintroduced. (Three-man flights had been halted in 1971, after Dobrovolsky, Volkov and Patsayev died in Soyuz 11 because they were not wearing their space suits.) They stayed for only four days, since the main objective of the mission was to test the new space vehicle.

On July 23, in the middle of the Moscow Olympic Games, Vietnamese guest Pham Tuan arrived at Salyut, followed on September 18 by Cuba's Arnaldo Tomayo Mendez. Both were greeted with the traditional Russian offering of bread and salt. After a brief stay, they returned to public receptions in their respective countries, celebrating the glories of Soviet communism.

While the preceding mission of Ryumin and Lyakhov had been complex and difficult, the new flight, as Ryumin and Popov set out to break the six-month barrier, went ahead in routine fashion, relieved only by the stories of the foreign visitors.

But on October 11, 1980, this stay, too, came to an end when Progress 11 docked with the station to bring supplies. The Dnieper crew returned to Earth in Soyuz 37, landing at the usual place, 180 kilometers from Dzhezkazgan. They had been in zero gravity for 185 days, but after landing, Ryumin and Popov left the capsule and walked by themselves to chairs waiting nearby. Their condition was therefore good, as was the condition of Salyut 6 in general. It had been designed for a two-year use but had already achieved three. Apart from the design, much of the credit went to the periodic maintenance to which the cosmonauts had dedicated 25 percent of their activities.

It was also clear that the sending of foreign guests under the heading "Intercosmos Mission" had become an institution. These visits, plus a series of short-duration all-Soviet missions, had settled into a rhythm of about one visit per month. Apart from the political motives, this arrangement also aimed at monitoring the physical and psychological condition of the crew members participating in long-duration stays.

PSYCHOLOGICAL PROBLEMS

Although Ryumin's second flight had passed with no serious problems, at least two episodes of crew behavior recorded in his diary hinted at a possible psychological deterioration. The first was on June 15, 1980, after two months in orbit. The treadmill the cosmonauts used for physical exercise was broken; but despite knowing how important its use was for their health both in orbit and on return to Earth, the cosmonauts had not repaired it, because "it meant undoing a lot of nuts and bolts and repairing it would take up a lot of time." When the doctors found out, they ordered an increase in physical activity, forcing the crew to fix the apparatus.

The second episode can be found in Ryumin's diary for September 10. The pair had now been in orbit for five months and had decided to forgo their monthly shower. Ryumin wrote: "When you start thinking of all the preparations to make and everything to be done afterwards, your desire for a shower weakens. You've got to heat the water, prepare the system with the protective sheet and arrange the water containers, attach the aspirators . . . well, it takes nearly a whole day just to have a shower."

Both episodes indicate a decline in the cosmonauts' awareness of their actions and their situation, leading to a change in their value judgments. This was certainly risky, given the restrictive safety margins of living in space.

THE STATION BEGINS TO DETERIORATE

Despite the generally satisfactory state of Salyut 6, its orbital age had already gone well beyond the set limits, and signs of wear and tear on the station were evident. Most problematic were the hydraulic systems for the climate-control apparatus, about which nothing much could be done. If they broke down, the station would be uninhabitable, and some bad signs were beginning to appear. To deal with these problems and in hopes of extending the base's life, a Soyuz T-3

flight was arranged and it departed on November 27, 1980, carrying Leonid Kizim, Oleg Makarov and Ghennady Strekalov. (Konstantin Feoktistov, a cosmonaut but, above all, one of the engineers responsible for the design of Soyuz, should have been in Strekalov's place, but the doctors had said no.)

The job of the new three-man crew, christened *Mayak* (Lighthouse), was principally to conduct a general internal and external review, replacing a series of equipment, especially electrical and electronic, and the pumps for circulating the cooling fluids in the station's external radiators. Then, they checked that no poisonous gases had stagnated in certain areas and verified the station's structural integrity. After this, Salyut 6 was again declared ready to host another

Cosmonaut Vladimir Dzhanibekov in Salyut 7.

long-duration mission as well as the scheduled visit of two foreigners. After providing the usual orbital boost, Progress 11 separated from the station on December 9 and broke up in the atmosphere. The next day, the crew returned to Earth after 13 days in orbit.

Notwithstanding these efforts, the year did not begin very well. After the arrival of Progress 12 with supplies and materials for a new crew, at the end of February, a solar-panel motor jammed and prevented its continuous pointing toward the Sun. This meant a power reduction of about 30 percent, so some heaters were switched off and internal temperature dropped to 10°C. This created a risky situation, because aside from the temperature being too low for comfort, condensation of water vapor gave rise to conditions liable to cause short circuits in the electrical wiring. The flight being prepared therefore had to take into account the new emergency and be ready to solve it.

On March 12, 1981, Soyuz T-4 departed, carrying Vladimir Kovalyenok and Viktor Savinykh (crew codenamed Photon). The third cosmonaut had been omitted on this occasion, and his seat was removed to make room for the repair equipment. The solar-panel motor was fixed immediately, followed by other elements of the climate-control system, and the internal environment returned to its normal, habitable state.

Ten days later, Soyuz 39 arrived as planned, bringing a Mongolian guest, Zhug-Derdemidyn Gurragcha, and the 60th anniversary of the Republic of Mongolia was celebrated. The series of visits concluded on May 15 with a Romanian, Dumitru Prunariu. Finally, Kovalyenok and Savinykh returned to Earth on May 26, ending the era of long-duration stays for the record-making station, Salyut 6.

THE TKS LOGISTICS MODULE AND THE END OF SALYUT 6

Although the manned program was considered finished, continuing use of the station for experiments not influenced by problems like climate control still held some surprises. Cosmos 1267 had been launched on April 25, 1981, and placed in a low orbit until the end of manned flights to Salyut 6. In the meantime, the Merkur capsule separated on May 24 and returned to Earth. Then, on June 6, Cosmos 1267 docked with the station permanently. The vehicle, which had the capacity to be manned, was known as "space vehicle for logistical transport" (Russian acronym TKS). The TKS project had distant military origins and indeed had been born under Vladimir Chelomei in the second half of the 1970s. Consisting of an FGB habitable module and a Merkur capsule, it had been tested with varying success on at least eight Cosmos launches before the flight to Salyut 6.

With the addition of the 15-tonne TKS, the cosmic convoy attained 32 tonnes. Its engines were fired twice to boost the orbit, taking the convoy to an average altitude of 350 kilometers. Finally, on July 29, 1982, the engines were fired once more to guide the fall of the

The shower in Salyut.

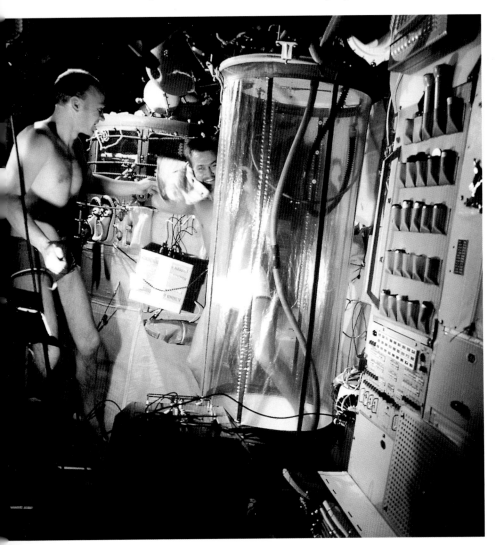

entire TKS/Salyut 6 space assembly over the Pacific Ocean, thereby decreeing its demise by destruction in the atmosphere.

The specialized modules for the future Mir station would later be derived from the TKS.

SALYUT 6: A RECAP OF SUCCESSES

In the light of experiences with the first five Salyut stations, nobody would have imagined the ultimate success record of Salyut 6. It had been active for 4 years and 10 months instead of the planned 2 years. During its orbital life, it had been manned for 684 days by 33 cosmonauts, 10 of them on long-duration missions, ending with Popov and Ryumin's record stay of six months. A further eight cosmonauts had been foreign guests from communist countries.

The engineers could rightly be proud of the cosmic home they had built, because routine maintenance had allowed it to be used safely. Indeed, only one serious breakdown (a fuel leak in one of the tanks) had occurred to threaten danger. But the repair had itself become an element of success, permitting activities to continue. In total, 36 vehicles had docked with the station, and only Soyuz 33 had failed to reach the orbital base because of engine failure. Two new vehicles, moreover, had been used for the first time: the manned Soyuz T and the Progress automatic cargo carrier. The latter's 12 flights had ensured the necessary supplies as well as the transport of 1 tonne of scientific instruments and maintenance equipment.

Overall, 1,600 scientific experiments had been carried out on the station using 150 different instruments, and 500 kilograms of scientific results had been brought back to Earth by Soyuz vehicles. The cosmonauts had made 60 astronomical observations, taking 13,000 multispectrum photographs, and had conducted 200 technological and 900 medical and biological experiments.

SALYUT 7: THE LAST OF A FAMOUS FAMILY

The year 1982 was a time of change at the Kremlin. Leonid Brezhnev, after being in power for 18 years, died in November. He had always supported a commitment to the space program, albeit less emphati-

Cosmonaut Oleg Atkov at the Salyut 7 control panel.

cally than his predecessor Khrushchev. Among other things, in 1961, he had been awarded the important title of Hero of Soviet Labor, together with Marshal Dmitri Ustinov and academician Mstislav Keldysh, for his political role in the early space conquests, in particular the development of the first rocket (also the first military intercontinental missile) that launched the first satellite, Sputnik, and the first cosmonaut, Yuri Gagarin. His successor was Yuri Andropov, who had previously acquired a certain notoriety as head of the KGB secret service.

On April 1982, while Salyut 6 was still orbiting unmanned with the mysterious Cosmos 1267 module attached, a new station, Salyut 7, departed from Baikonur. Consequently, for three months, two Soviet bases occupied the same orbit inclined 51.6° to the equator. The cause, it was claimed later, was French astronaut Jean-Loup Chretien. Initially, he was to have stayed on Salyut 6, but instead, he inaugurated foreign visits to Salyut 7. In the meantime, the old base was being kept in reserve in case the new station could not accommodate Chretien. Thus, Salyut 6 remained active until the end of the French mission.

Salyut 7 was the last task in a long job begun 10 years earlier with Salyut 1, and it was hoped that continuous occupation could be achieved, without having to lose time turning the station off and on every time a crew left and another one arrived after a pause. That objective would not be met, but the final tally would be positive in any case, consolidating a certain routine of work initiated with Salyut 6.

In general, the new orbital base was the same as the preceding one, but it did have improvements to some systems aimed at increasing its reliability and reducing the work the cosmonauts had to do to manage the station. For example, the forward docking collar had been strengthened to absorb the impact of heavier vehicles more safely, and the solar panels had been built with the capability to add smaller ones on the sides to compensate for power loss caused by natural degradation of the solar cells. An ingenious but extraordinarily simple method had been thought of to aid bacterial cleaning of the internal atmosphere. The windows of two portholes were made so that they did not filter ultraviolet radiation. This radiation entering the cabin killed a certain percentage of the bacteria in circulation. A large porthole had also been added in the transfer chamber to facilitate astronomical observations, and all the equipment was protected on the outside by transparent antimeteorite covers when not in use.

Significant improvements had also been introduced to the flight system, using new versions of the Delta navigation system and the Kaskad attitude-control system that now functioned with an accuracy of 1°. The station had reached sufficient reliability to be maneuvered from Earth without supervision by a live-in crew. The smaller size of the new environmental systems ensuring survival on the base meant that five people, two crew members and three visitors, could be present at the same time, even for continuous periods.

Interesting changes had also been made to life onboard. There was continuous hot water, hot plates for cooking and a small refrigerator for fresh food delivered by the Progress automatic vehicles. All the food items had been separately wrapped so that meals could be made according to preference instead of having to choose from prepackaged combinations. Four meals were planned each day, for a total of 3,150 calories. Finally, water was no longer brought up in small, separate containers but stored in a large reservoir refilled directly by Progress vehicles. Unfortunately, problems still existed with the Bania shower and the razor for shaving. As in the past, various biomedical (Aelita, Polynom, Pinguin) and gymnastic (veloergometer, treadmill, etc.) equipment was available, but with improved performance.

To help keep frequently misplaced objects in place, the architects of the cosmic home had built into the walls a number of small lockers with washable surfaces. Equal care had been taken over the lighting, which had been revised to make it more efficient and more consistent in the various zones.

As far as fixed scientific equipment was concerned, the customary cone rising from the floor of the work area contained an X-ray telescope (RT-4M) and an X-ray spectrometer (SKR-2M). New versions of the two cameras (the multispectrum MKF-6M and the cartographic KATE-140) had been added to study Earth, together with two spectrometers for oceanographic studies, a Yelena-F gamma-ray spectrometer to study radiation near Earth, and Oasis and Phyton greenhouses for experiments in plant cultivation. Later, a Progress vehicle would deliver Korund, a second-generation furnace (superseding the previous Splav and Kristall), inside of which it was possible to reach 1,270°C for experiments in producing new alloys and semiconductor crystals on an industrial scale.

The goal of Salyut 7's designers was to use it fully and continuously for almost four years. This was achieved, but then the station remained unmanned in orbit for another five years. During its use, however, it accommodated four short-duration missions of 8 to 12 days and six medium- (50 to 65 days) and long-duration missions, extending the record to 237 days (nearly eight months).

Cosmonaut Svetlana Savitskaya during a Salyut 7 space walk.

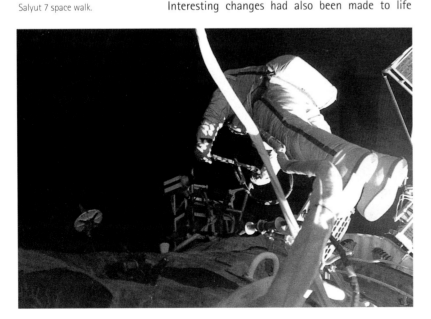

SIX MISSIONS: PROBLEMS AND MANY VISITORS

The first crew to enter the new station consisted of two cosmonauts: Anatoli Berezovoi and Valentin Lebedev. The latter recounted his adventures after returning to Earth by publishing a diary of his seven-month (212 days, to be precise) stay in orbit.

The launch took place from Baikonur on May 13, 1982, with Soyuz T-5, and the rendezvous should have been facilitated by a new electronic instrument named Mera, which was theoretically capable of locating the station when it was still 250 kilometers away and guiding Soyuz to its destination more quickly and easily. But the instrument "discovered" the base when it was only 30 kilometers away—that is, just before Salyut's Igla system began working toward the same purpose and completed the operation with the help of an Argon computer.

The station was activated, and the new occupation began, to the joy of Russian ham radio operators, because a small satellite named Iskra that they had built

was launched from it (via the hatch usually used to jettison refuse, which then burned up in the atmosphere).

Everything functioned normally on the station, but the cosmonauts faced a reproof from the controllers at Kaliningrad when the Progress automatic capsule arrived with the first supplies. Attracted by the possibility of watching the vehicle dock through a rear porthole, the cosmonauts stayed in the tunnel leading to the work area. According to regulations, however, the tunnel should have been closed and inaccessible during the maneuver. The admonishment did not worry them unduly, however, and they started to transfer material as though nothing had happened— "while I listened to Progress creak metallically," wrote Lebedev. Progress's engines were used to boost the orbit, and then it departed.

When the cosmonauts had to have their first monthly shower, they did so unwillingly, like their predecessors, because of the lengthiness of the operation. In the meantime, an inspection revealed that the

Cosmonauts Vladimir Dzhanibekov (right) and Viktor Savinykh (front) inside Salyut 7.

portholes without ultraviolet protection had dark marks at the edges, perhaps caused, the experts said, by radiation acting on the rubber sealant.

On June 25, after a month in orbit, it was time for the first visit of another guest—Jean-Loup Chretien, a French air force test pilot, who had the honor of launching another ham radio operators' Iskra satellite from the usual waste hatch. During his week onboard, Chretien was mainly occupied with medical experiments, while the crew did work in astronomy and materials science, including making a new aluminum alloy in zero gravity.

Chretien spent an intense eight days trying to complete all the scheduled experiments. Still, he was satisfied, as Lebedev wrote in his diary, adding that the French cosmonaut found the station "simple, nothing special but reliable." There were some exceptions, Lebedev continued, recounting the night he had to get up to go to the bathroom and found the toilet full of urine. He had to work for an hour to empty it before finally being able to use it.

Life onboard and the experiments conducted generated refuse. "Drops of coffee and juice, adhesive tape and scraps of food floated in the air," wrote Lebedev. "So once a week, we did the cleaning, using detergent on the walls, the table, portholes and instruments and using a vacuum cleaner in the lockers." Two events in July and August made the crew happier. On July 30, the two cosmonauts exited for a space walk to change some scientific instruments and to try out functions like turning nuts with special pliers. The excursion lasted only 25 minutes, but the sight of Earth and the black sky dotted with stars was, as ever, an emotional experience. When they returned, Lebedev saw that the protective covering of his helmet had been sliced, probably by some protuberance on the station, but he had not noticed when it happened. "Thank God the helmet was built with a double layer of metal," he wrote in his diary.

The second event was the visit of a female colleague, Svetlana Savitskaya, accompanied by Alexander Serebrov and Leonid Popov. With light brown hair and dark eyes, Svetlana had just passed her 34th birthday and had a university degree in engineering.

Before going into space, she had gained some fame in the communist world as an aerobatic pilot and chaser of aeronautical records. Lebedev and Berezovoi welcomed her with a bouquet of flowers grown in the space greenhouse, and they had prepared one of the areas in the Soyuz capsule for her to have a little more privacy. Although she appreciated the gesture, Svetlana preferred to rest with everyone else in the sleeping bags hung on the walls. The company of the second Soviet woman cosmonaut, following in the footsteps of the legendary Valentina Tereshkova, lasted just a week.

Meanwhile, Lebedev and Berezovoi continued their mission, including enacting a simulated emergency escape from the station. Fortunately, that need never arose. December 10 arrived, and it was time to return to Earth. "I looked around before leaving the station," wrote Lebedev, "and it all seemed so familiar, everything was close and dear to me. I felt no alienation because of the isolation; it had become my home."

But the last days had not been very happy. Berezovoi was not feeling well, and the Kaliningrad controllers suggested preparing to go home. Then the Delta automatic navigation system began to cut out frequently, making work more complicated. In any case, they passed the 211th day, and another barrier was broken. However, their landing was one of the worst, taking place in a snowstorm that prevented helicopters from reaching the capsule for several hours.

THE SECOND CREW AND A SERIES OF ACCIDENTS

Three months went by, and Salyut 7 orbited the Earth empty and alone until an automatic vehicle, Cosmos 1443 (a habitable module like the one that had joined Salyut 6), arrived and used its engines to lower the orbit to 300 kilometers. So preparations were made for the arrival of a new crew. But the Soyuz T-8, which was launched from Baikonur on April 22, failed to reach the station because the rendezvous antenna was damaged during the ascent, and the capsule was forced to return.

Things went better for the next crew, consisting of Vladimir Lyakhov and Alexander Alexandrov, who were launched on June 27 in Soyuz T-9. They were to

remain in space for 149 days (nearly five months). First, they went into the Cosmos module to retrieve items stored there, including the new Delta navigation system. It was the first time cosmonauts had entered this new type of module, with its Merkur capsule. The latter was filled with material to be taken back to Earth, then Cosmos de-docked but remained in orbit until August 23, when Merkur separated from it and reentered the atmosphere.

A nasty, unexpected accident occurred at the beginning of September during refueling with Progress 17. A conduit for corrosive nitrogen tetroxide inside the engine chamber ruptured, rendering part of the engine block useless. The engines on the still attached Progress were used to shift altitude.

A new crew was therefore prepared to deal with the problems in orbit. But when Vladimir Titov and Ghennady Strekalov blasted off from Baikonur on September 26, their mission almost ended in tragedy. The two cosmonauts were saved by the emergency system

that ejected Soyuz and its crew away from the rocket shortly before it exploded.

Work carried on in Salyut, and around the end of October and the beginning of November, the two cosmonauts went outside to install new solar cells around the central panel and to increase the power supply by 50 percent. Then they returned to Earth.

THE THIRD CREW AND REPAIRS

Now the station had to be repaired, and this required the right tools and training. The former arrived with a Progress carrier, while a crew able to deal with the problem lifted off in Soyuz T-10B on February 18. Onboard were Leonid Kizim, Vladimir Solovyov and a medical doctor, Oleg Atkov.

After they entered the darkened station and reactivated it, there was a brief visit by a crew with a foreign guest, Rakesh Sharma from India. Then they prepared for space walks to repair the fault. This was Kizim and Solovyov's job, but the work proved to be

Cosmonaut Vladimir
Dzhanibekov inside Salyut 7.

Illustration of docking between the Cosmos 1443 module and Salyut 7. Facing page, Salyut 7 in orbit.

complicated. Work was suspended after four excursions (totaling 14 hours and 45 minutes) in April and May, and months passed before the next attempt.

The next space walk was dedicated to adding another solar-cell panel. Another crew came to visit on July 18—Vladimir Dzhanibekov, Igor Volk and Svetlana Savitskaya once again. She and Dzhanibekov carried out 3.5 hours of extravehicular activity, performing cutting and welding tests on small specimens of metal.

Only on the sixth space walk, in August, was the repair finally completed, thanks to a hand-held pneumatic press that had been brought up by the last crew (who had already gone home). Despite this, the station's engines still were not used.

It was clear that things were not going well on Salyut 7, and on October 2, 1984, Kizim, Solovyov and Atkov returned. The situation remained unclear, and when links between the control center and the base were completely cut off in February, the daily newspaper *Pravda* made the sibylline announcement that the station had been "deactivated, continuing the flight in automatic mode."

CRISES AND A FOURTH CREW

On June 6, 1985, Soyuz T-13 lifted off from Baikonur carrying Viktor Savinykh and Vladimir Dzhanibekov, who had visited the station the year before. As they approached with the aid of a laser telemeter, they saw that the base was completely lifeless, and for the first time, Soyuz used only its own resources to dock with the large, dead body hanging in space. The disastrous situation was confirmed after berthing, when Soyuz' instruments revealed that there was no power on the station. The two cosmonauts opened the hatch and went inside carefully: the interior was freezing, perhaps –10°C, but the air was breathable, although it stank of mold. Fortunately, no noxious gases had formed.

After dressing in heavier clothing, the crew began their inspection and found that a defect in an orientation sensor had sent the base careening out of control so that the solar panels were no longer pointing at the Sun. The eight batteries had gradually run down, and two were completely useless. One after the other, instruments had shut down, including radio links, meaning that the control center had no way of know-

ing what was happening. Kaliningrad had therefore watched helplessly as Salyut died without an explanation.

Using Soyuz' engines, Dzhanibekov and Savinykh pointed the station toward the Sun, and the solar panels began to generate power and recharge the surviving batteries. At the same time, Soyuz also controlled the air inside. It would take a month, until the end of July, for the internal climate-control system to return to normal. It took this long because they had to wait for the ice to vaporize in the frozen areas so that water did not get into the instruments and destroy them. In any case, the ice had managed to damage the water-heating system, and a powerful lamp normally used for television filming had to be employed to supply heat.

In spite of everything, the station was almost miraculously reborn thanks to the work of the cosmonauts and the ground technicians. The last tasks were completed after the arrival of Progress 24, which delivered spare parts, three batteries, fuel and solar cells for the third panel that had not yet been extended. This last operation was carried out by the cosmonauts on August 2 during a five-hour space walk in new, semi-rigid space suits also delivered by Progress 24.

In the meantime, another vehicle, Cosmos 1669, docked briefly with the station. It was later discovered that this was a new type of Progress that would eventually be used with the Mir station and was therefore being tested.

THE FIFTH CREW

Once the major repairs were over, the crew was ready to return. But before that, a new crew—Vladimir Vasyutin, Alexander Volkov and Georgi Grechko—arrived on September 17. It was the first time a crew exchange had taken place without an interval. A few days later, on September 25, Dzhanibekov and Grechko went home, leaving behind Savinykh, who became part of the new crew that was to remain onboard for 65 days. Actually, a much longer stay, until the end of March 1986, had originally been planned. But in October, Vasyutin felt ill, with aches and a high temperature. Kaliningrad attempted a diagnosis, but it was impossible. The situation worsened to such an extent that it required the immediate interruption of the mission and a return with Soyuz T-14 on November 21. Vasyutin was taken to hospital in Moscow, where it was discovered that all his problems were the result of a prostate infection.

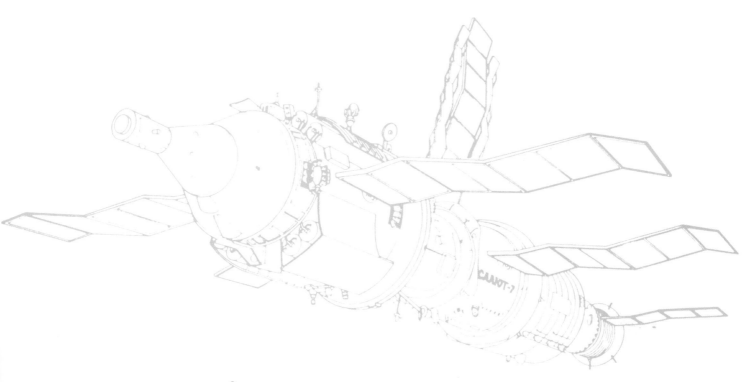

Meanwhile, on October 2, Cosmos 1686 (another habitable module in the TKS series, with a Merkur capsule) had arrived at the station bringing 4.5 tonnes of supplies and new equipment, all of which remained aloft unused. There was talk of sending another traditional crew as well as an all-woman crew. Svetlana Savitskaya, Yekaterina Ivanova and Yelena Dobrokvashina were prepared for the latter. But time passed, and two events intervened. The first was that Savitskaya became pregnant and was no longer able to go into orbit, so the all-woman flight was canceled. The second was the launch of a new station, *Mir* (Peace/Union), on February 20, 1986.

THE LAST CREW

The last crew to say goodbye to Salyut 7 would also be the first to enter Mir, establishing an almost perfect continuity between the Russian cosmic homes. On March 13, 1986, Leonid Kizim and Vladimir Solovyov took off, headed for Mir, where they stayed for 50 days. Then the two cosmonauts got back into Soyuz T-15, in which they had arrived, and moved to Salyut 7, where they stayed for another 51 days.

Two space walks (3.5 hours and 5 hours) were made during the mission to raise a mobile tower, then retract it and finally extend it again. Welding tests were also carried out on some joints of the tower. Salyut 7 was finally abandoned on June 25, 1986, and the crew returned to Mir.

At that point, the Salyut 7/Cosmos 1686 assembly used the latter's engines to boost the station's orbit to 475 kilometers to prevent rapid orbital deterioration. In this way, it survived another five years until February 7, 1991, when it plunged into the atmosphere and disintegrated over South America. Various pieces were found on the ground in Argentina.

Undoubtedly, Salyut 7 had had a varied and difficult orbital life compared with Salyut 6. All the same, the end result of 2,500 scientific experiments carried out by the cosmonauts over 813 days of occupation was

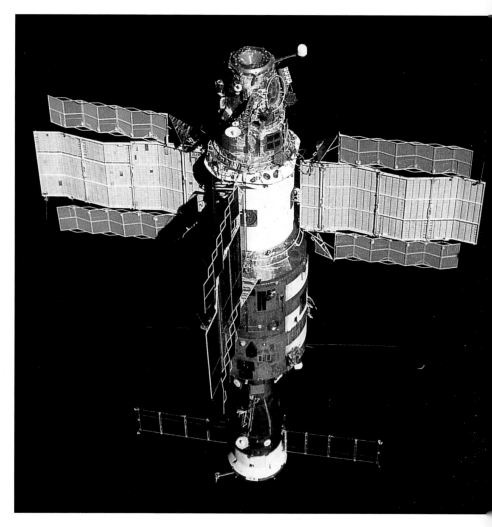

Salyut 7 in orbit.

judged to be positive and significant. And from the point of view of technology and engineering, even the problems had offered opportunities to test working methods and means that considerably enriched the know-how necessary for the next and more difficult step forward. That step was a bigger station—like Mir—that would be inhabited for even longer periods.

The family of seven Salyuts had served that end. Although the last two second-generation stations had provided greater possibilities for use, they were still experimental stations whose primary job had been to try out technologies that could then be routinely applied on Mir. Mir was, in contrast, considered an "operational orbital base."

The odyssey of the Russian space station Mir

4

Six living and
working modules
and many accidents

The odyssey of the Russian space station Mir

On February 20, 1986, there was a surprise launch. In the early-morning hours, a Proton rocket took off from Baikonur to carry the space station *Mir* (a name meaning "Peace" and also "New World") into orbit. No one was expecting a new space base, because Salyut 7 was still orbiting the Earth with Cosmos 1686 attached (increasing its habitability and potential). It had been empty since November 21, when the crew had returned home hurriedly because one of the three cosmonauts, Vasyutin, was sick. Work had been only suspended, however; everything on Salyut was still functioning.

Nonetheless, a new, third generation of stations commenced on the eve of the February 24 opening of the 27th Congress of the Communist Party of the Soviet Union. From the platform of honor, Mikhail Gorbachev, in power for only a few months, spoke of the need for radical change after the stagnation of the Brezhnev years. All the same, the old tradition of linking a space event to a political one was evidently still held to be valid.

The year 1986 left its mark for good and bad on Moscow. After embarking on the new policy that was to lead to the collapse of the Soviet empire, Gorbachev met with Ronald Reagan in Reykjavik in October. In December, he freed dissident physicist Andrei Sacharov, who had been in exile at Gorky since 1980. But in the months before that, Gorbachev had had to face the gravest nuclear accident in the history of atomic technology. The core of the Chernobyl nuclear reactor in the Ukraine melted down, spreading radioactive pollution across the continents.

While the USSR used Mir to take another leap forward in human settlement of space, the United States was paralyzed after the disaster of the Challenger shuttle that exploded shortly after launch on January 28. Little solace came from the accomplishments of NASA's Voyager 2 probe four days earlier, when it flew past the surface of Uranus, the third-last planet of the solar system, at a distance of only 81,000 kilometers.

The Soviet space program was now in the last great phase of its history. In 1986, the Vega probes flew by Halley's comet, along with Europe's Giotto, and work at Baikonur was progressing toward the launch of the world's biggest rocket, Energia, and the Buran shuttle. In the end, both of the latter proved to be swan songs. No one doubted, though, that Mir was a revolution in space stations.

Preceding pages: Mir in orbit.

SIX SPECIALIZED MODULES

The hopes for Salyut 7 were fulfilled with Mir: the station had six modules specialized for different activities, and it enabled continuous occupation by a crew, with no gaps. For the first time, therefore, the space base would always be operational.

Ultimately, it would take 10 years to assemble it in orbit completely, and the collapse of the Soviet Union would heavily influence the duration of its life in space. Instead of the planned five or six years, it was still in service in February 1998, the 12th year, with the prospect of remaining so for at least another year and a half. The events that characterized its long tenancy above the Earth colored the history of the conquest of space.

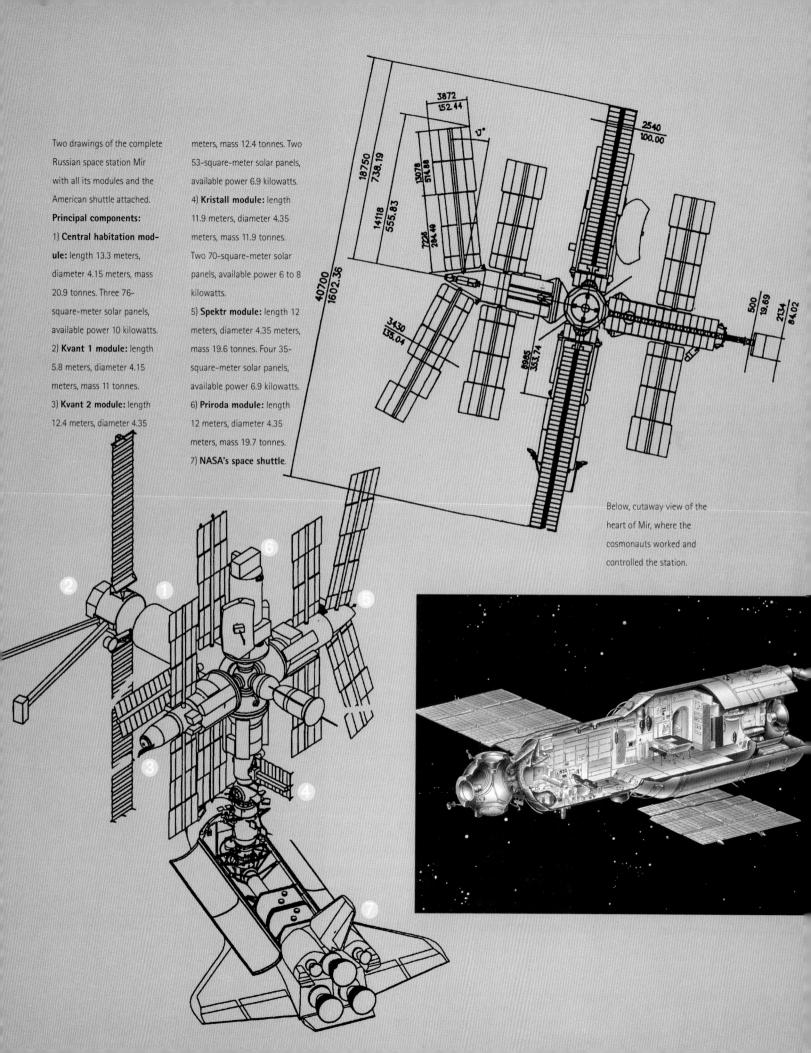

Two drawings of the complete Russian space station Mir with all its modules and the American shuttle attached.

Principal components:

1) **Central habitation module:** length 13.3 meters, diameter 4.15 meters, mass 20.9 tonnes. Three 76-square-meter solar panels, available power 10 kilowatts.

2) **Kvant 1 module:** length 5.8 meters, diameter 4.15 meters, mass 11 tonnes.

3) **Kvant 2 module:** length 12.4 meters, diameter 4.35 meters, mass 12.4 tonnes. Two 53-square-meter solar panels, available power 6.9 kilowatts.

4) **Kristall module:** length 11.9 meters, diameter 4.35 meters, mass 11.9 tonnes. Two 70-square-meter solar panels, available power 6 to 8 kilowatts.

5) **Spektr module:** length 12 meters, diameter 4.35 meters, mass 19.6 tonnes. Four 35-square-meter solar panels, available power 6.9 kilowatts.

6) **Priroda module:** length 12 meters, diameter 4.35 meters, mass 19.7 tonnes.

7) **NASA's space shuttle.**

Below, cutaway view of the heart of Mir, where the cosmonauts worked and controlled the station.

Above, three of the Russian ships deployed around the world to act as relay stations for communicating with Mir. Above right, Mir during construction on the ground. Facing page, three phases in the growth of Mir. From top to bottom: the first, basic module where the cosmonauts lived; the station with Kvant 1 and a Soyuz capsule docked at the rear; Kvant 2 and Kristall modules attached.

When the first Mir module took off on February 20, some people were not only surprised but also disappointed. Some had imagined a new type of station with a different design from its predecessors. But Mir's structure was the same as Salyut's: the same diameter and more or less the same length. The only difference was a spherical forward unit with five docking systems (four lateral and one axial), which replaced the tapered cylinder of Salyut's only forward docking port. A sixth docking port was added at the rear to be used by the Progress automatic vehicles. All this left no doubt that other modules were expected. Noting the lack of sizable scientific equipment inside, malicious critics pointed out the station's poverty. But the first module was the station's command center, built to accommodate a permanent crew of two cosmonauts plus guest crews of three people staying for short periods.

The heart of the base. Inside, the traditional, cumbersome upturned cone that rose from the floor of Salyut to expose observational instruments to the vacuum of space had disappeared. In its place was a table for five people extending from a side wall, with a hot-water tap to reconstitute dehydrated food. The walls, full of storage drawers, were color-coded by section and function, as suggested by the psychologists for identification and also to provide stable, conventional reference points. The narrower section housing the command and control center had a dark green floor and light green walls; the larger area had a brown floor and yellowish walls. The ceiling, bearing lights, was white throughout.

Personal cabins and bathroom. Toward the back, two cabins, one on each side, were for the permanent residents of the base. For the first time, an area had been designed to give the cosmonauts a bit of privacy. Each cabin had a berth on one wall, a folding table and a porthole to view the Earth, the stars and the darkness. It was all pretty spartan, but the curtain had only to be drawn so that the cabin became isolated from the rest of the station. The temporary guests had the usual sleeping bags hung on the ceiling and walls.

To keep the interior free, the veloergometer (exercise cycle) could be folded into the floor, while the treadmill remained. In the same area, behind the right-hand cabin, there was a bathroom, elegantly called "sick bay" by the cosmonauts, a punning term indicating that life above was not always easy, as sometimes the body rebelled against such unnatural conditions. Inside, besides the toilet, there was also a transparent sphere with rubber inserts for washing face and hands.

Structure: the walls of a space base. Mir's walls were made of chemically treated aluminum of varying thickness: 2 millimeters in the living area, 1.2 millimeters in the narrower command area and 5 millimeters in the sphere with the docking systems. On the exterior were the radiators and the heat shields made of 25 layers of aluminized Mylar covered by other layers of a material similar to Kevlar. Together with the internal climate-control system, this was the only way to deal with the continual 300° alternations in temperature to which the base was subjected, going from −170°C when the station was in shade to 120° above when it was lit by the Sun.

The internal atmosphere. The internal temperature could vary between 16° and 28°C, and the humidity could also be high enough to condense into vapor for transfer to the water reserves. The internal atmosphere was Earthlike, with oxygen and nitrogen supplied separately from tanks and then mixed. Oxygen was extracted from water by an Elektron electrolytic system. The hydrogen left over from electrolysis was expelled into space. In emergencies, 40 cartridges of sodium chlorate could produce oxygen when heated to high temperatures (around 800°C). The cartridges were sometimes used when there were a lot of people onboard and consequently an increased demand for oxygen.

Internal pressure was the same as at sea level. Carbon dioxide produced by breathing was removed by passing the air through zeolite containers that absorbed the gas and returned clean air. There were also emergency absorbent cartridges in case the primary system failed. Other substances were used to remove any dangerous trace gases (some produced by the crew), like methane, carbon monoxide, hydrogen and ammonia.

Water and recycling. Water was available in three types: for drinking, for washing and that obtained from electrolysis. Three purification systems were onboard. One obtained drinking water from condensed humidity and was similar to the one used on the last two Salyuts. Water used for hygiene was recycled for the same purpose. Water was extracted from urine and then electrolyzed to produce oxygen for breathing. The urine-treatment equipment was housed in two modules named Kvant 1 and 2. In total, 60 percent of the water used onboard was recycled.

Power. The necessary power came from gallium arsenide solar-cell panels that had been tried on previous stations. They were now the norm and increased the available energy by 30 percent in comparison with silicon. The heart of the station had two large lateral solar panels, and arrangements had been made to insert a third, vertical one during a space walk to make available a total of 9 to 10 kilowatts. Power was stored in 12 nickel-cadmium batteries. All the panels were kept pointing at the Sun. The other modules also had solar panels: four on the one named Spektr and two each on Kvant 2 and Kristall. Altogether, they supplied another 30 kilowatts.

Navigation and guidance "brain." A set of seven electronic computers, called Strela, installed on the station was based on the Argon systems used on the Soyuz capsules. Two were the "central computers." The one with the higher capacity (500,000 operations a second) was an Argon 16 with increased memory, used mainly for position control. The second was a standard Argon (200,000 operations a second) like the

one on the Soyuz TM and was used as a reserve. Two cathode-ray monitors on the command console displayed the data being handled. Among the other important computers for station management, one was dedicated to the navigation systems and replaced the old Delta used on Salyuts 6 and 7.

The station had a Kurs system for the forward docking unit, reserved for Soyuz capsules. This used a radar transponder able to lock onto the capsules when they were still 180 kilometers away. At the rear, where the Progress automatic vehicles arrived, the older Igla from the 1970s was used. This came into action at only 25 kilometers. However, when the Kvant astrophysics

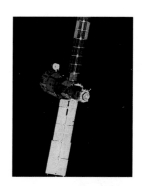

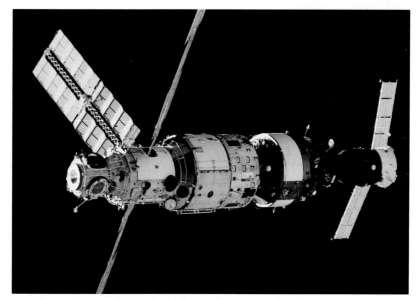

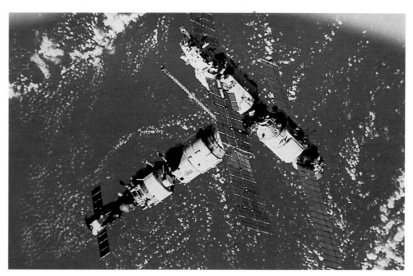

Launch of a Semiorka rocket from Baikonur carrying a crew for the space base.

module was permanently attached to the rear, its docking apparatus used both the Igla and Kurs systems.

As far as position control during flight was concerned, the optimum station attitude was with the longitudinal axis perpendicular to the Earth's surface. Bearings were taken with a series of optical (sextant, solar and stellar sensors) and magnetic (magnetometer) instruments, infrared sensors, accelerometers and gyroscopes that could determine orientation very accurately (to within 1.5° in normal conditions). To control attitude, the computers responded to specific commands mainly by activating the Gyrodynes, six large gyroscopes (three for each axis, doubled for safety) whose movement by electric motors caused a change in the station's attitude. If these were not enough, a propulsion system of 32 small rockets (16 plus 16 reserves) positioned around the station could be used as an alternative. Each booster could provide a thrust of 14 kilograms. Use of the Gyrodynes was preferred, though—first, because they did not pollute the surrounding environment and create problems for the scientific observations, and second, because it was better to save fuel.

Telecommunications. Mir was linked with six Luch (also known as Altair) relay satellites for data and voice telecommunications. The satellites were in a 36,000-kilometer, geosynchronous, equatorial orbit and provided a constant link with the control center in Kaliningrad (now Korolev). Previously, when Russian ground stations and ships at sea equipped with retransmission antennas had been used, the maximum available linkage time had been only 20 minutes every 90-minute orbit, when the station passed over Russia.

Daily routine and meals. A typical weekday began at 8 in the morning (and ended at 11 in the evening, when it was time for bed). After two hours for personal hygiene and breakfast, the crew worked from 10 a.m. to 1 p.m. Then an hour of exercise, one hour for lunch and three more of work were followed by another hour's physical activity. After dinner, the evening was free. Exercise on a treadmill (running 5 kilometers) and a cycle (10 kilometers) could be done at intervals—it was important to do about two hours a day to burn off about 450 calories.

Stations in science fiction: *Star Trek*

Gene Roddenberry's TV series Star Trek *began in the 1960s. In 1979, Paramount released* Star Trek—the Motion Picture *(directed by Robert Wise), in which the crew of the starship* Enterprise, *commanded by Admiral Kirk with assistance from the Vulcan Spock, had to defend the Earth from a mysterious alien force.*

The daily diet contained 100 grams of protein, 130 grams of fat, 330 grams of carbohydrates, plus various vitamins and minerals. The majority of the food had to be rehydrated, and some could be heated on the stove provided. There was also a refrigerator for fresh food. Refuse was collected in special bags. When these were full, they were jettisoned through an airlock—an exit compartment—also used to expose scientific equipment to the vacuum of space. Other waste was placed in the Progress supply vehicles before these de-docked from Mir at the end of their mission and then disintegrated on reentry.

Drugs and alcohol. Some drugs were available onboard for first aid in the case of space sickness and minor disorders like headaches. There were also tranquilizers and cardiotonic drugs. Alcohol was officially banned, but it seems to have later been permitted on certain occasions.

ARRIVAL OF THE LABORATORY MODULES

Kvant 1, an aft module. A little more than a year had passed since the launch of the heart of the Mir station when the first laboratory module, Kvant 1 (for astrophysics), left from Baikonur on March 31, 1987. (Kvant 1 was the forward part of a TKS module, already tested on the Salyuts. It was designed so that after docking with Mir, its rear engine section would separate from it.) However, the attempt to dock with the aft unit on April 5 failed. First, the target was missed, and the module passed within 10 meters of the base. Then on April 9, another attempt was only partially successful, as something prevented a firm connection between the two components.

The cause could not be explained. Cosmonauts Yuri Romanenko and Alexander Laveikin, who had been onboard Mir for nearly a month, had to go outside to make a direct inspection. They found that a waste-disposal bag had jammed between the parts, preventing closure. Laveikin pulled it out with his hands, and complete docking was at last possible.

Kvant 2, asymmetrical architecture. Another three years or so were needed before the base could continue its planned growth with the addition of a second module, Kvant 2. Launched on a Proton from Baikonur on November 26, 1989, its experience was as hair-raising as that of its predecessor. Once it was in orbit, the Kaliningrad control center discovered that one of its two solar panels had not opened completely. To fix this would require the difficult double maneuver of rotating the module and the panel at the same time. However, on December 2, when Kvant 2 approached Mir's forward docking port, the Kurs control system stopped it 20 meters away because it was approaching too fast. At the same time, the station's computer had deactivated the attitude-control system because of an error signaled by the Gyrodynes.

Once again, it was intervention by the cosmonauts that resolved all the problems. Alexander Viktorenko and Alexander Serebrov had been living on the station since the beginning of September, and the main purpose of their mission was to receive and activate the new module, which originally should have left on October 16 but had been delayed because of a problem with the onboard computers.

Viktorenko and Serebrov therefore took control of the critical situation created by the double breakdown and manually piloted the station to complete the docking. Afterwards, the Lyappa robotic arm was used to transfer Kvant 2 to a lateral docking port to free the central one.

The second module had a hatch for space walks as well as systems for position control and life support. It

Arrival of a Soyuz capsule ready to dock with Mir.

Above, Mir complete with all six habitable modules and two Soyuz capsules docked at the rear.

Below, the small greenhouse; and a cosmonaut's morning hygiene.

had two oxygen generators: Elektron plus the Vika cartridges. This reduced the need for hauling supplies of the precious gas by Progress vehicles.

Kvant 2 contained various pieces of scientific equipment ranging from an incubator for the in-orbit birth of small birds to chambers for multispectrum photography. Two circular openings in one wall allowed terrestrial resources to be studied.

Kristall, geometry restored. There was less of a wait for the arrival of the third laboratory module, named Kristall. Its launch was postponed several times but finally took place on May 31, 1990. Kristall, too, seemed dogged by the same bad luck as its predecessors. On June 6, docking with the station was aborted because of a failure in the attitude-control motor, which ran longer than planned, causing the operation to be halted. Four days later, another attempt succeeded with no further problems. After docking, Kristall was moved to the side opposite Kvant 2 by the robotic arm, restoring a balanced, T-shaped geometry.

Kristall was practically the same as Kvant 2 except for its spherical end section, where two androgynous docking systems were to connect with the Buran shuttle that was then in development. The Kristall

module specialized in production of new materials like semiconductor crystals for microelectronic components. For this purpose, it had five furnaces (Krater, Optizon, Zona 2, Zona 3 and Kristallisator), plus two solar panels that could be removed and transferred to another part of the base, if necessary.

Two modules in storage. At this point, the new Soviet space home had reached a good size, but its management was proving to be more difficult than expected. For example, the complexity of replacing the Argon computers with the more powerful Salyut 5B version had required more time than planned, which delayed the start of a series of scientific experiments. In theory, a larger crew was needed, but the base's systems permitted this for only brief periods. So the project managers decided to suspend delivery of the last two modules, Spektr and Priroda, respectively slated for 1991 and 1992. They were already being built in the Khrunichev workshops, but both were put into storage.

The Sofora tower rises from the station. One of the various tasks assigned to the cosmonauts was to try out construction techniques. It immediately became clear that this was not a simple job when four space walks (on July 15, 19, 23 and 27), totaling 24 hours' work by Anatoli Artsebarski and Sergei Krikalyev, were needed to erect Sofora. This trellis structure, with a 1.5-meter-square base, was a sort of 14.5-meter-high tower made of tubular metal poles and anchored to the Kvant 1 module. After its construction, a metal Soviet flag bearing the hammer and sickle was added to the top. Later on, in September of the following year, a group of motors would be fitted to the top of Sofora to improve the station's attitude-control capability.

FALL OF THE USSR: ENLARGE OR ABANDON MIR?

In 1989, Valentin P. Glushko's position as head of the Korolev Center (later renamed NPO Energia) was taken by Yuri Semyonov, long a protégé of the powerful Andrei Kirilenko. On December 25, 1991, President Mikhail Gorbachev resigned; the USSR ceased to exist, and flags and symbols were changed. The red hammer-and-sickle flag remained only on Mir's Sofora tower, and it would be removed only in September 1992, during a space walk to install the motors at the top of the tower.

However, the future of the station had already been called into question during the coup d'état in August 1991. Krikalyev was still onboard, but his companion Artsebarski had been replaced on October 4 by Volkov, accompanied by Austria's Franz Vieboeck and Toktar Aubakirov from Kazakhstan. The latter had been included in the crew at the last moment to improve relations with his newly independent country, which was home to the Baikonur center. However, that meant leaving behind the Russian cosmonaut who should have relieved Krikalyev (who agreed to stay on in any case).

The Austrian could not be grounded, because $7 million had been received for his participation in the mission, and with financial problems already emerging, that money was precious. The economic situation was such that cutbacks had even been made by terminating the flight-assistance activities of the fleet of Soviet ships deployed in the oceans. Furthermore, the collapse of the USSR brought closure of some centers in countries now separated from the former Soviet Union that had been receiving signals from the two Luch relay satellites. The result was that in February 1992, the station was out of communication with the Kaliningrad control center for nine hours a day.

Birth of the new Russian space agency: The new president, Boris Yeltsin, set up the Russian Space Agency (RKA), directed by Yuri Koptev, to run all civilian space activities, including those of the Mir station (previously under control of the ministry for machine construction, in practice an arm of the defense ministry).

At this point, the station was still incomplete but had already reached the planned five-to-six-year limit of its technical life span. And indeed, problems of deterioration in some of the systems began to appear at the end of 1991, when 5 of the 12 large gyroscopes used on the Kvant modules failed. As no resources were left to send the new Mir 2 station into orbit as planned in 1992, there was now the problem of what to do with Mir 1: abandon it or prolong its life with the necessary maintenance. In spring 1992, B. Chertok, the head of NPO Energia (the center that supervised the running of Mir), announced that the

Scenes from life on Mir. From top to bottom: Doctor Valery Poliakov examining European astronaut Ulf Merbold; America's Shannon Lucid with Onufrienko (left) and Usachyov (background) at the lunch table; the crowded central area of Mir, with Solovyov (center) and Bonnie Dunbar (lower left).

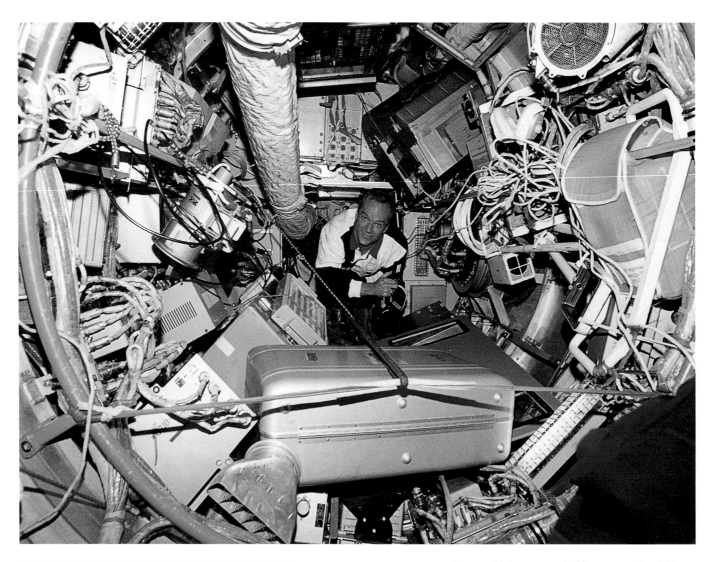

station would be put on hold, unmanned, within a couple of months. This did not happen, but his statement showed just how uncertain the future was.

THE "PANORAMA" INSPECTION

In spite of everything, activities continued, albeit at a slower pace and by extending ongoing missions (as with Krikalyev's). In the meantime, the operation of checking and replacing the deteriorating systems began. Apart from the Gyrodynes, the air-conditioning unit and some parts of the systems for communicating with the Luch satellites were replaced in spring 1993. Then in the fall, three space walks by Vasily Tsibliev and Alexander Serebrov were dedicated to an external inspection of the shields (according to a plan

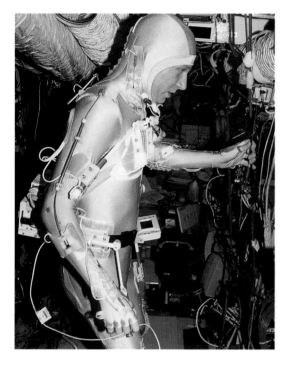

named Panorama). A 5-millimeter hole was discovered in a solar panel, probably made by a meteorite. This may have happened the previous August, when the Perseid meteor shower had been 2,000 times more intense than usual and microcraters 1.5 to 4 millimeters in diameter had been gouged into the portholes.

But at the end of the Panorama inspection, the station's external state was declared to be in good shape, as only tiny, surface impacts had been found, none of which had penetrated the base's structure to create risks.

ELEVEN U.S.-RUSSIA MISSIONS

Mir's fate became clear only in September 1993, during the first meeting of the U.S.-Russia mission for economic and technological cooperation, when American Vice President Al Gore and Russian Premier Chernomyrdin decided to extend the mutual commitment to manned space flight. That agreement, which would later make Russia a partner in the new space station, also produced a plan for close collaboration on Mir for two years, including 11 flights of the U.S. shuttle to the Russian station, stays on Mir by NASA astronauts and scientific programs involving the two modules (Priroda and Spektr) that had not yet been launched but were almost ready. NASA financed the plan by pro-

viding $400 million to the Russian space agency, $330 million of which was to modernize the base.

In February 1994, Krikalyev was the first Russian cosmonaut to fly on the American space shuttle Discovery. In February 1995, it was Discovery, with Vladimir Titov among its crew, that made the first shuttle rendezvous with the station.

DOCKING OF SPEKTR, THE FOURTH MODULE

Soon afterwards, on May 20, 1995, the Spektr module was launched; it docked with Mir without difficulty 10 days later. It was then berthed opposite the Kvant 2 module (after moving Kristall), and even the small problem of failed solar-panel deployment was resolved.

The new piece of the station was similar in size and characteristics to the others. It housed spectrometric, radar and TV equipment for research on terrestrial resources and to study the Earth's atmosphere, ecology and meteorology. It also had X-ray and gamma-ray detectors to analyze the interactions of radiation with the Earth's magnetic field. Part of the instrumentation (750 kilograms) was American-made.

The interior of Spektr was the same as Kvant 2 and Kristall, although it was slightly shorter because the rear wall was closed and did not have an exit hatch or an attachment for other capsules; outside the tail was a set of equipment for environmental observation.

Besides being a laboratory, Spektr was also to be living quarters for a visiting NASA astronaut and so would contain his personal effects, from clothes to toothbrush. The first to inhabit it was Norman Thagard, who arrived in Mir on June 25, 1995, after the first, historic docking between the American shuttle and the Russian station. They were joined with a compartment anchored in the shuttle's cargo bay. Berthing took place with the Kristall module temporarily transferred to the front of the station to keep it safer during the operation and to allow freer movement with no structures of the base in the vicinity.

NASA had prepared a "connecting tunnel" that would

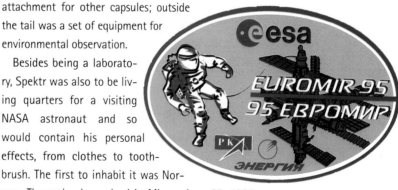

Facing page: top, American astronaut Charles Precourt during a visit to Mir; cosmonaut Yuri Usachyov in his personal cabin, bottom. Left, ESA astronaut Thomas Reiter during a medical test.

Cosmonaut patch from the ESA's Euromir 95 mission.

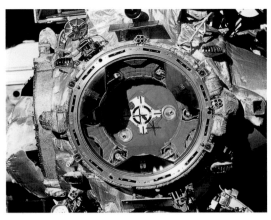

be used for the next joint mission in November, and this was left as a permanent and integral part of Mir. Then separate visits by six American astronauts began.

THE PRIRODA MODULE COMPLETES THE STATION

On April 23, 1996, the last module, Priroda, took off, and three days later, it joined the orbital base, thus completing its complex architecture after 10 years.

Priroda was equipped with a variety of instruments (about 1 tonne of which was American) to investigate the terrestrial environment, detect emissions of cosmic radiation and conduct experiments in making new types of materials.

THE EUROMIR MISSIONS

The European Space Agency (ESA) joined the cooperative Mir program, aimed toward construction of an international space station. Through the Euromir project, two astronauts, Ulf Merbold and Thomas Reiter, both from Germany, would stay on the Russian base.

The first stayed for 31 days in 1994, and the second for 179 days in 1995/96, acquiring experience in long-duration missions that ESA astronauts were lacking up till then.

Merbold and Reiter carried out various types of research using instruments prepared by European scientists for biomedicine, technology and the treatment of materials. During shuttle-Mir missions, ESA also tested an electronic station-rendezvous system. One of Reiter's experiments, named ESEF, measured meteorite impacts. This revealed that during the period the instrument was exposed to space, the station passed through a stream of almost invisible debris twice a day that produced 5,000 microscopic impacts in just one minute. This information proved to be very useful in designing protective shields for the International Space Station.

BREAKDOWNS AND ACCIDENTS

The station's condition remained fine and problem-free until 1991, when 5 of the 12 large Gyrodyne gyroscopes

in the Kvant modules broke down. That same year, problems began to appear in the computers too. Britain's Helen Sharman recounted that during her stay in May 1991, the computer controlling the station's position malfunctioned so that the station was no longer pointed at the Sun, electrical power failed and everything onboard slowly went out as the batteries ran down.

Other elements of the various systems (environmental control, communications) then had to be replaced. But at the beginning of the 1990s, the Russian space agency was aware of the situation and managed the station in a different way, based on ongoing maintenance that became an integral part of the cosmonauts' daily life. At some stages, in fact, they had to spend as much as two hours a day fixing their cosmic home.

In November 1995, however, a more serious breakdown occurred. One of the Vozdukh systems installed on the Kvant 1 module to remove carbon dioxide failed. Lithium hydroxide cartridges were used to absorb the gas, but there were enough cartridges for only 30 days. Then it was discovered that the problem was due to a loss of liquid refrigerant, which threw the environmental control system off balance and made the carbon dioxide removal system inefficient. The leak was immediately stopped, and the proper tools and spare parts were prepared to be brought up a few days later on the space shuttle Atlantis, which took off on

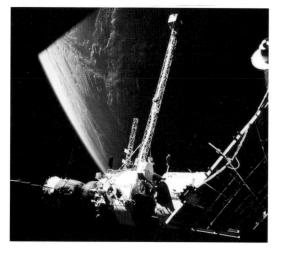

Top left, cosmonaut Valery Poliakov looking out of a porthole on Mir. Above, the American shuttle approaching the station. Left, the Sofora tower erected on the Kvant 1 module, with rocket engines at the end.

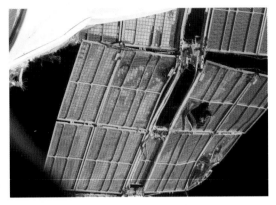

Top, two photographs of the Spektr module's solar panel that was damaged on June 25, 1997, in a collision with the Progress M-34 automatic vehicle.

Above, the new hatch for the Spektr module brought from Earth and fitted with electrical connections to the module's remaining undamaged solar panels.

November 12. As a precaution, the opportunity was also taken to increase the reserves of cartridges.

Furthermore, every now and again, there were problems in the electronic systems during rendezvous and docking maneuvers. Sometimes, as on August 27, 1994, with Progress M-24, the difficulties were overcome by the intervention of the cosmonauts, who completed berthing manually. The risk was mainly that an automatic Progress vehicle might go out of control and come very close to the station; this would become dangerous as safety margins for the cosmonauts onboard were reduced to nearly nil.

In addition to the dangers of these "close encounters" were the satellites that crossed Mir's orbit or flew close to the station. As Vera Medvedova, spokesperson for the Korolev Center, noted, every month, a couple of orbiting objects passed in the station's vicinity (though usually outside the 1,500-meter safety limit), and on two cases, an emergency situation occurred. On November 8, 1993, the 6-tonne Russian satellite Cosmos 1508, launched 10 years earlier and now dead and wandering, passed within 300 meters of Mir. On September 15, 1997, an American spy satellite came dangerously close, and Solovyov, Vinograd and Foale were ordered to put on their space suits and take refuge in the Soyuz capsule, ready for departure in case of trouble. The satellite passed within 470 meters of the orbiting base, which, to compound the risk, was out of control because its computers had failed and were not completely repaired.

1997, A BLACK YEAR

The most serious accidents in Mir's history happened in 1997. The two greatest dangers astronauts can face

on a station are fire and collision with another vehicle. The first will not only quickly damage vital parts of the base but also cause smoke that can affect the environmental control system and turn the cabin into a gas chamber. A collision, on the other hand, could result in violent depressurization and a horrible death for the astronauts. Both things happened in 1997, but fortunately without becoming a catastrophe.

The fire. On February 24, the crew—commander Vasily Tsibliev, Alexander Lazutkin and the American, Jerry Linenger—were using the Vika cartridges (sodium chlorate cartridges that produce oxygen when heated to nearly 800°C) to supplement the oxygen in the Kvant 1 module. But an internal oxygen leak got into an electrical circuit and started a fire. The crew used fire extinguishers immediately, but the flames, although contained, went on for almost 10 minutes. The crew wore gas masks for a few hours and breathed oxygen from tanks while waiting for the smoke to disperse and the air exchange system to do its job. Linenger, a doctor, checked his two companions and saw that they were all right. For the next few days, however, they all wore light masks so as not to breathe the particles still floating in the air.

A little over a week later, on March 5, the Elektron system in the Kvant 2 module that supplied oxygen by electrolysis broke down. A similar system in Kvant 1 had been out of use for some time. All the crew could do now was use the cartridges left in Kvant 2, because the ones in Kvant 1 had been destroyed in the fire. Each man used one cartridge a day, and there were enough reserves to reach the beginning of April, when a Progress supply vehicle was due to dock.

But what if it was late or something happened on its trip? Then the station would have to be abandoned. With this prospect hanging over the station, which would mean the probable end of Mir, another accident on March 19 worsened the situation. A directional sensor failed, and the computer turned off systems and put the station into emergency orientation, with the four modules facing downwards (like a downturned flower). A few days were needed to activate the reserve sensor and restore a proper balance.

In the end, on April 8, Progress M-24 docked with Mir, full of spare parts and 60 new oxygen cartridges. About a month later, the shuttle Atlantis also arrived, bringing a new Elektron system to replace the one on Kvant 1, which had been out of service for some time, plus a number of lithium hydroxide cartridges to remove carbon dioxide. Linenger returned to Earth as arranged, and his place was taken by Michael Foale.

The collision. Meanwhile, another serious accident was on its way. After having been filled with refuse, Progress M-34 was de-docked. Then, on June 24, to test the new TORU remote-piloting system, Commander Tsibliev at his command post guided Progress as if he were onboard it to re-dock with the station. But he lost control of Progress, which struck the Spektr module, damaging a solar panel and the cooling radiators. The collision with the panel strained its connection with the station, opening a crack through which air escaped.

The specter of depressurization hung over the crew. Tsibliev and Lazutkin, realizing that the leak was in the Spektr module, ordered Foale to shut himself in the Soyuz capsule. Then they quickly closed the access hatch and isolated Spektr from the rest of the station. However, this meant a considerable cut in power, because Spektr's four panels supplied nearly half the total needs.

The problem of depressurization had been overcome (pressure had dropped by only 15 percent once the hatch was shut), only to be replaced by the question of power. With grave consequences. All the instrumentation had to be shut down. But keeping the station turned on and in the correct position required power, particularly for the large gyroscopes. It was decided to turn these off to save the batteries, putting the station in an inertial mode with the heavier modules underneath. When Mir could again communicate with Kaliningrad, the controllers ordered that the Soyuz capsule be turned on to use its engines to position the station with its panels pointing at the Sun.

It took some days for the batteries to recharge completely, and in the meantime, the crew had to work with flashlights and without environmental systems. It was cold onboard, air exchange was inadequate, and even the toilet could not be used much because the

The Orlan space suits and the rocket pack

Space walks outside Mir were made by cosmonauts wearing an Orlan DMA space suit, good for 10 six-hour EVAs. Weighing 105 kilograms, the suit held 2 kilograms of oxygen and 3.6 kilograms of water. It worked on batteries and with an umbilical power cable connected to the station. It was donned via an opening at the back of the suit.

In February 1990, the rocket pack, far right, was also used. Nicknamed the "space motorbike," it could take cosmonauts up to 50 meters away from the station with an umbilical cord attached and up to 100 meters without. The Russian rocket pack weighed 400 kilograms and was equipped with 32 small nitrogen steering rockets. It could be used for six hours in semiautomatic or completely manual flight moving at a maximum speed of a few meters per second.

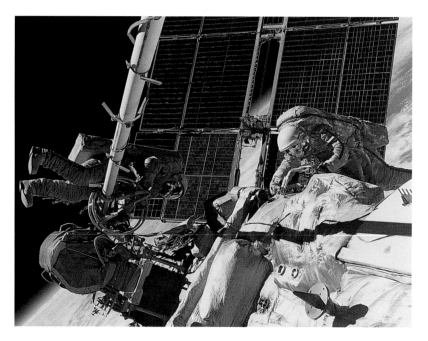

Above, the space walk by cosmonauts Yuri Gidzenko and Thomas Reiter. Bottom right, scientific instruments installed on the Kvant 2 module. Facing page, Mir in orbit.

electrolytic recycling system was turned off. So there was little oxygen, and cartridges had to be used while the crew wore masks and kept the extinguishers at hand in case of another fire. This went on for a few days, while the Progress M-35 supply vehicle, scheduled to leave on June 27, was kept on the ground, waiting to send up the tools necessary for repairs.

It was decided to send up a hatch, already prepared to connect on one side with the power cables in Spektr and on the other with those in the docking compartment. In this way, the link would be restored, while the hatch would remain closed and the module isolated.

The cosmonauts would have to put on their extra-vehicular activity (EVA) space suits, isolate all the modules of the station and enter the depressurized Spektr to attach the plugs. In the simulator on the ground, cosmonauts Pavel Vinograd and Anatoli Solovyov successfully checked out the maneuver. Then on July 5, Progress M-35 was launched with the necessary parts.

Heart problems. On July 12, commander Tsibliev's heartbeat showed worrying anomalies caused by stress. The doctors ordered him to stop all activity and take medication. His place at the repairs should have been filled by the American, Foale. Unfortunately, however, while they were preparing the operation, Lazutkin accidentally unplugged the computers, and the station lost orientation again. This revealed the cosmonauts' critical psychological and physical state, and although

another time-consuming repositioning maneuver was needed, the controllers in Kaliningrad ordered their return. Repairs were postponed until the arrival of a new crew, Vinograd and Solovyov, on August 7.

Upon their return to Earth on August 14, Tsibliev and Lazutkin—the two cosmonauts who had lived through the "black period" of Mir, having been on the station since February 12—were subjected to an enquiry that established, according to Valery Ryumin (the Russian coordinator of the Mir/NASA program), their responsibility for the serious accidents and recommended they be severely fined. A little later, however, the vice president of the Russian space agency, Boris Ostroumov, announced that the government would be asked to decorate them. President Yeltsin calmed the waters and defended the two cosmonauts, explaining that "as always in Russia, a scapegoat was sought and the easiest thing was to accuse the crew."

On August 22, Vinograd and Solovyov, dressed in their space suits, finally conducted the long-delayed "intravehicular activity"—that is, they entered the Spektr module, which had been closed since June, to reconnect 11 cables through the new hatch brought from Earth. It was dark inside, and some parts were covered in ice. But everything appeared to be in order, and a little more than three hours later, the operation was concluded.

Malfunctioning computers and other surprises. More problems occurred, mainly created by the computers. These had malfunctioned in the past, but from

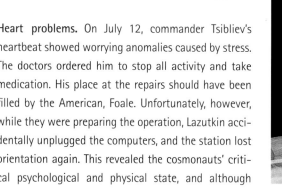

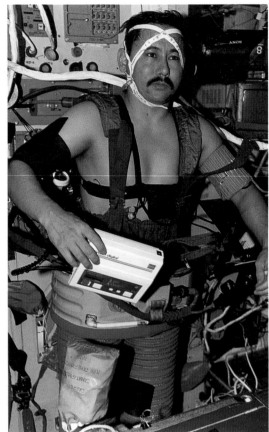

August on, they became one of the major periodic sources of trouble. Spare parts were sent up from Earth, but they never seemed sufficient.

Other problems derived from the two Elektron oxygen-producing systems. One had been out of order since the end of June and remained so until the fall.

Moreover, in early November, at the end of a space walk, the hatch of the Kvant 2 module where the cosmonauts exited did not form a perfectly hermetic seal, and from then on, a slight loss of pressure was evident. It took another three months to discover the cause and fix it. Then in December, there was a leak in the liquid refrigerant, and the air-conditioning system failed, so the one on Soyuz had to be used.

Even the mission of the tiny Inspector robot vehicle (93 centimeters long and weighing 72 kilograms), built by DASA for the German space agency, was a failure. Released from the Progress M-36 vehicle 500 meters away from the station, it should have circumnavigated the base to make an inspection using its TV

cameras. Instead, the cosmonauts lost control of it, and it was moved farther away to avoid a collision. But despite these seemingly endless problems, the crews continued to work.

The year 1998, too, seemed to open badly, as on the second day of the year, the cosmonauts again had to deal with malfunctioning computers. In February, smoke began issuing from an instrument, but this was immediately controlled and the instrument was turned off.

All the same, these incidents paled into insignificance after those of the previous year. Meanwhile, the big repair of the crack at the base of the solar panel on the Spektr module still had to be dealt with. At the beginning of March, a space walk—to set up a handrail and support points in the repair area—that had already been postponed several times was interrupted, because the pliers the cosmonauts used to open the exit hatch on the Kvant 2 module broke. Everything was put on hold once more.

THE INHABITANTS OF MIR

From February 1986 until the first quarter of 1998, around 80 cosmonauts and astronauts stayed on Mir, some more than once. The total with visitors is in fact higher and reaches more than 120, including the American shuttle crews who, for a few days, passed between the station and the docked shuttle. A crowd, therefore, left its marks in Mir's cabins, which were increasingly filled with wires, tools, tubes and various materials. The Institute of Space Microbiology in Moscow revealed that the 94 families of microorganisms found on the station in 1990 had grown to 140 in 1997; they were divided into 54 types of bacteria and 86 fungi. Some had even settled in the portholes. It was obviously the mixed, international community of occupants who had brought these organisms aboard.

While the majority of Mir's inhabitants were Russian, the list of non-Russians was also long and varied. Besides the group of cosmonauts sent up for political reasons (Abdul Mohmand from Afghanistan, Bulgaria's Alexander Alexandrov, Mohamed Faris from Syria and Toktar Aubakirov from Kazakhstan), there were also the first paying astronauts. These included Austria's Franz Vieboeck, Japan's Toheiro Akiyama of the Tokyo Broadcasting System (the first journalist in space), Helen Sharman from the UK, Germany's Klaus-Dietrich Flade and Reinhold Ewald and the French group of Jean-Loup Chrétien, Michel Tognini, Jean Paul Haignere, Claudie André-Deshays and Leopold Eyhard.

To these were added the ESA and NASA astronauts involved in medium- and long-duration flights as training for the new International Space Station. The Europeans were Ulf Merbold and Thomas Reiter, respectively involved in missions known as Euromir 94 (31 days) and Euromir 95 (179 days).

All told, there were seven Americans. Starting in June 1995, the first to stay onboard was Norman Thagard (115 days), followed by Shannon Lucid (setting a women's record of 188 days), John Blaha (128 days), Jerry Linenger (131 days), Mike Foale (143 days), David Wolfe (128 days) and, finally, Andy Thomas (128 days), who returned to Earth on June 12, 1998.

Apart from Commander Vasily Tsibliev, who hit the station with the Progress vehicle on June 25, 1997, two other very famous Russian cosmonauts took part in the life of Mir. Yelena Kondakova stayed on the base for 169 days in 1995 (setting a Russian women's record as of 1998). Valery Poliakov, a doctor by training, stayed on Mir for two lengthy periods. The first started in November 1988 and lasted 240 days. Then he left Earth again in January 1994 and stayed for nearly a year and a half (437 days, to be exact), the longest period spent by a human in orbit (again as of 1998).

Thagard and Lucid, the first Americans, did not tolerate the isolation very well and had some difficulty adapting. But they were not the only ones. Kondakova, too, was quite irritable, and Poliakov said that sometimes, she even refused to eat.

Space advertising. On a couple of occasions, the cosmonauts agreed to appear in commercial advertisements. During a space walk in May 1996, cosmonauts Onufrienko and Usachyov inflated a 1.2-meter cylindrical replica of a Pepsi-Cola can as the operation was filmed for a commercial. In August 1997, despite the crisis onboard, Lazutkin made a commercial for the Israeli milk company Tnuva. It cost $450,000.

THE TWILIGHT OF MIR

In July 1998, Russian government sources and representatives of the RKA space agency announced that Mir's reentry into the atmosphere would take place in July or August 1999. But NPO Energy, the association responsible for Mir, found an American investor and so, upon the completion of what would have been its last mission in August 1999, another one began in April 2000 with the new prospect of being able to continue.

Above, Yuri Semyonov, head of NPO Energia, the organization responsible for Mir. Left, the station simulator used to prepare the cosmonauts at Star City, near Moscow.

5

Freedom, the "paper station"

Reagan's decision:
an undertaking with
friendly nations

Freedom, the "paper station"

The success of the first American station, Skylab, in 1973/74 led NASA to reconsider the possibility of beginning a program to construct a real orbiting station. Despite everything, Skylab had never truly been a space station, because it was a result of adapting the third stage of the Saturn V rocket and thus did not have resupply systems, one of the fundamental requirements for prolonged use. Although the American space administration had begun allocating most of its resources toward building the space shuttle, in the mid-1970s, research was recommenced to determine the best course of action. To that end, in 1974, NASA administrator James Fletcher requested a survey "to identify and examine the various possibilities for a civilian space program for the next 25 years."

A REPORT TO BEGIN WITH

The resulting report ("Outlook for Space," published in January 1976) considered the prevalent lack of enthusiasm for space and the lively ongoing debates on the limits of development. In that light, the report concluded: "The future programs must guarantee a public service . . . contributing to the development of new energy sources, answering the new environmental challenges, aiding the prediction of natural or manmade disasters and improving the production and distribution of foodstuffs." At the time, critics claimed that it was better to invest public money in meeting the immediate needs of the population rather than in taking on costly enterprises beyond Earth. So NASA attempted to use its document to ensure its survival and get closer to the requirements of the times. If this was valid in principle, it was not in everyday practice, because space activities still had to develop the necessary technological capabilities before being really useful for concrete and terrestrial ends.

Thus, the report was not particularly enthusiastic about the idea of a space station, despite considering it "the next logical step" after the shuttle. It was not seen as "an end in itself, but rather as the technological support for a variety of other objectives that could benefit from the increase in knowledge accrued by man's capacity to work in space, and laying the foundations for future development."

THE MOSC PROJECT

The station studies heeded these principles and, in spite of everything, tried to satisfy them. An example was the Manned Orbital Systems Concepts (MOSC) project drawn up in 1975 by NASA's Marshall center together with the McDonnell Douglas Astronautics Company. This covered a "scientific base for scientific and technological research for programs directly linked to improving life on Earth." All the same, the project, which outlined a modular, expandable station orbiting

Preceding pages: illustration of space station Freedom and the shuttle.
NASA Marshall Space Flight Center design for the MOSC station.

Left, two illustrations of the
NASA Marshall Space Flight
Center's SASP station design.
Below, the shuttle with the
large solar generator to
prolong its stay in orbit.

at 360 kilometers, also examined other activities to do with the construction of large structures and installations dedicated to space production. The manned modules were derived from Spacelab, built by Europe's ESA for NASA's shuttle, and were to be capable of housing a crew of four astronauts to be changed every 90 days. Cost of the operation: $1.2 billion.

But in the fall of 1975, NASA changed tack. The concept of an orbiting laboratory was no longer enough. Instead, preference was given to a "base of operations" that could be used not only for research activities but also for the assembly, launch and control of space vehicles. Support would also be given to activities in geostationary orbit at 36,000 kilometers above Earth, where the geostationary telecommunications and meteorology satellites already operated.

Moreover, these years also saw a big push toward futuristic projects led by two visionary scientists. Gerard O'Neill at Princeton University was analyzing (as far as possible) details of the concept of large orbiting colonies, and Peter Glaser of the Arthur D. Little company was working on Solar Power Satellites, very large satellites able to collect solar energy and beam it to Earth in the form of microwaves. These notions actually had little effect on the American space adminis-

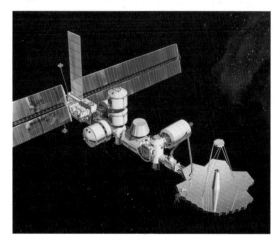

tration, but they did serve to rekindle prospects that had seemed exhausted.

So in December 1975, NASA invited private-sector proposals, and in April of the following year, it awarded two study contracts—Space Station Systems Analysis Studies (SSSAS)—to McDonnell Douglas and Grumman Aerospace, which were to work under the respective direction of the Johnson Space Center in Houston and the Marshall Space Flight Center in Huntsville. The work of both teams was aimed at designing systems to collect solar energy, as well as studying the terrestrial environment, telecommunications, commer-

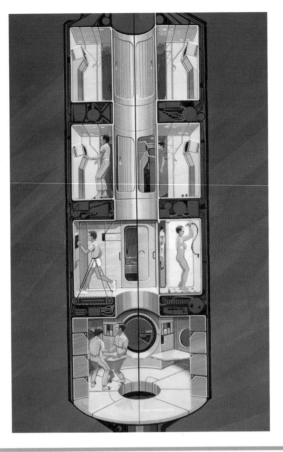

Right, Johnson Space Center's early 1980s plan for the interior of a manned module. Above, illustration of the SOC station concept.

cial production of materials in zero gravity and general scientific research.

McDonnell Douglas kept to its previous basic concept of an expandable base and indicated 1988 as the year of completion for a cost of $2.4 billion. Grumman concentrated more on the collection of solar radiation and the automated construction of orbiting structures, estimating that everything could be ready for 1986.

THE SHUTTLE FOR THE STATION

In contrast to these highly advanced projects, there was a current of prudent conservatism at NASA trying to impose a more immediate, cheaper, less complicated approach. The proposal that seemed to have the upper hand for a time was a simple extension of the space shuttle's capabilities beyond the 15-to-30-day mission limit. Research was done on structures which could be left in orbit and to which the shuttle could dock, mainly to increase its power resources using large solar panels.

This "economical" theory involved platforms, larger than traditional satellites, to be left in space for vari-

Stations in science fiction: *Alien*

Ridley Scott's 1979 film Alien *made a star of its leading actress, Sigourney Weaver. It is the story of a monster that invades the spaceship Nostromo by taking over the body of an astronaut. The film's success at the box office led to a number of sequels in the following years.*

ous automated activities. These were to be used principally for Earth observation with various instruments and some scientific research. A common characteristic was a power module made up of large solar panels whose record size brought the platforms to more than 150 meters.

The tide of interest influenced Europe, where the ESA was beginning construction of the Eureca platform, a smaller and simpler design that ultimately was to have little fortune, being used in orbit only once. The almost identical SPAS platform built by the German company MBB (later Daimler Aerospace) was used repeatedly on shuttle missions, and in the United States, the Fairchild company was trying to develop something similar.

THE SASP AND SOC PROJECTS

The initial studies for the SSSAS program were followed by others involving more corporations. Two markedly different plans resulted. The Marshall center, with McDonnell Douglas and TRW, proposed the Science and Applications Space Platform (SASP), which was really a station to be built with existing technologies and dedicated mainly to scientific applications. However, it could also be manned. This was an evolution of previous McDonnell Douglas studies involving the concept of modules, the first of which generated power. The base would then grow, increasing the possibilities for crew accommodation from an original two people. The project was later broadened, under the name Evolutionary Space Station, to include the possibility for in-orbit construction, storing supplies, inter-orbital vehicle transfers and acting as a "base of operations."

It was the late 1970s, and the SASP project incorporated all of the specifications determined to date. It was also a response to the rival project drawn up by the Johnson Space Center and the Boeing company that proposed a real Space Operations Center (SOC) orbiting the Earth. The Johnson center in Houston wanted to use this to relaunch its role in manned missions. It based its plans on the supposition that many space activities would in future be carried out in a 36,000-kilometer geostationary orbit.

The space station would therefore act mainly as a support base in geostationary orbit. Consequently, the idea was to build ferry vehicles able to shuttle between the station and satellites working at a much higher altitude. On the SOC, structures and satellites would be built; it would also provide support and maintenance for manned spacecraft, with the idea of transferring numerous formerly Earth-based activities to the station—where they could be done at a lower cost, according to the proposers.

Unlike the Marshall-McDonnell Douglas project, the SOC tended virtually to exclude research and scientific objectives, "tolerating" at most the temporary docking of a module if there were enough ports free. In practice, the designers' philosophy was that scientific research should be done on free-flying platforms near the station, to be visited only when results had to be collected. The SOC, for which Boeing was awarded a research contract in July 1980, would also take advantage of the development of new technologies. Its cost was estimated at $9 billion. Finally, its initial capacity to house four astronauts in 1990 would increase to 12 or even 20 by 2000.

Illustration of the SOC station design by NASA's Johnson Space Center.

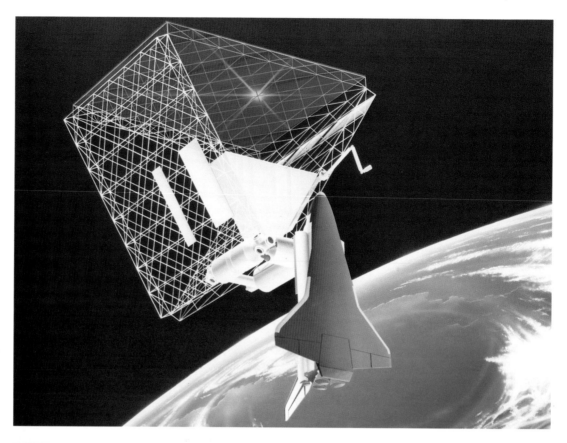

Right, one of the six station designs considered—the Triangular station.

Below, meeting between U.S. President Reagan, British Prime Minister Thatcher and Japanese Prime Minister Nakasone during the Economic Summit held in London in June 1984. The meeting also decided the participation of international partners in the space-station project. Britain was later to withdraw, however.

THE "NEXT LOGICAL STEP"

Meanwhile, work on the reusable space shuttle was completed, and flights began in 1981. NASA now had to make a strategic decision on a new program, and a space station at last seemed to be a real "next logical step." Station projects had already been at the point of starting in the late 1950s and again in the late 1960s, only to be shelved in favor of first the Moon landing and then construction of the shuttle. But now, there seemed to be no obstacles on the horizon. In addition, since the 1960s, NASA had carried out so many station-concept studies that it now had enough knowledge to move on to actual design and construction.

In May 1982, NASA administrator James Beggs set up a Space Station Task Force to prepare a reasonable and feasible plan to submit for the White House approval necessary for all strategic commitments. "It's easy enough to design a space station," said John Hodge, chairman of the task force, "but it's difficult to put together all the elements into a plan that is useful for the country and realistic from the point of view of current economic conditions."

The task force therefore set up three working groups to determine the station's mission, its characteristics and the planning of its construction. A fourth group was later added to determine the engineering and budget aspects. In the end, a station design emerged that would carry out eight basic functions, incorporating all those examined in the various projects up till then. At one and the same time, it had to serve these functions: a laboratory for scientific and technological research; an orbiting observatory for the Earth and space; a node for space transport; a base for servicing and maintaining satellites, platforms and ferries; a space production center; a vehicle-assembly workshop; a storehouse for fuel and spare parts, etc.; and an experimental base for future, more ambitious projects concerning lunar colonies and a landing on Mars. After years of waiting, nothing was left out, and the message seemed to be that there was always time to delete something.

It was agreed, however, that construction would be evolutionary and that the eight specified functions were objectives to be achieved gradually. To that end, the architecture had to be flexible to allow gradual expansion, and maintenance and resupply became two other cardinal elements to ensure long-term survival.

Finally, the cost of the enterprise was fixed at $8 billion (in 1984 dollars). While the financial commitment for design and development started with the 1987 fiscal year, projected completion of a usable station was in the early 1990s. Naturally, use of the base would be open to scientific and commercial operations on an international scale.

PRESIDENT REAGAN LAUNCHES
THE STATION PROJECT

In 1981, Ronald Reagan replaced Jimmy Carter in the White House, and in June of the following year, the president approved a National Space Policy document that laid down guidelines for the country and therefore NASA. Among other things, these emphasized strengthening international cooperation, commitment to scientific and technological research and commer-

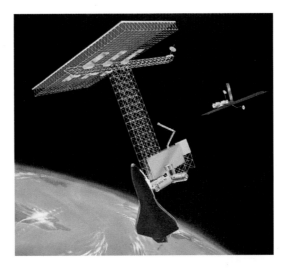

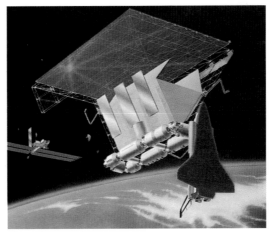

Three of the designs considered for the space station: Streamlined, left; Triangular, bottom left; and Raft, below.

cial uses of outer space, as well as repeating that the space effort must ensure continued U.S. international leadership. The context into which the space station was to be born had been defined.

However, to determine the basis on which the president would make his decision, the Senior Interagency Group for Space (made up of representatives from the departments of state, defense, trade and transport, the directors of the Central Intelligence Agency, the organization of the chiefs of staff, the armaments control and disarmament agency, the NASA administrator and the heads of the management and budget office and the scientific and technological policy office) requested a review of NASA's proposal and its alternatives.

Subsequently, on April 7, 1983, the station project proposal was presented to the president, and on April 11, he signed the Senior Interagency Group's request for further review of the matter, with an eye to drawing up a final document by September. This did not happen, though, because the group failed to reach a consensus on the presidential directives, and other ways and means were therefore used to provide the White House with the information needed to make its decision.

Meanwhile, concern was growing among scientists that commitment in this direction would require so many resources as to cause cancellation of other pro-

grams. This was the same apprehension as had been expressed at the beginning of the shuttle program 10 years earlier. The president's scientific advisor, Dr. George Keyworth, suggested that any station project be included in a longer-term plan which included the building of lunar colonies. The Congressional Office of Technology Assessment, an influential body in Congress, also shared this opinion. Consensus was slowly weakening, or even disintegrating, for a political decision on the apparently imminent station. After so many years of technical, scientific and social assessment, now that the moment for a definitive decision was at hand, it seemed almost as if no interest in building the station existed. Or at least, the voices of those against it prevailed.

Meanwhile in the Soviet Union, as Yuri Andropov replaced Leonid Brezhnev, the successful era of Salyut 6 was ending as the station reentered the atmosphere in July 1982, after five years of frenetic activity and a continual stream of foreign visitors. Before that, in April, Salyut 7 was launched to consolidate an unarguable supremacy in manned orbital bases. The United

States, on the other hand, had not progressed past Skylab, now 10 years old and without a follow-up. Their only manned module, the shuttle's Spacelab, was born in Europe under the ESA.

In 1983, President Reagan had thrown down a challenge to the USSR by initiating the SDI strategic defense initiative, the so-called "Star Wars." That challenge now continued on a civilian level, and notwithstanding negative advice and criticism, on January 25, 1984, Reagan gave the go-ahead to construction of the space station in his State of the Union Address to the combined Houses.

"This evening," the president announced, "I am asking NASA to develop a permanently manned space station and to do it within the decade." He added: "We would like our friends to help us in this endeavor and to share the benefits. NASA will invite other nations to take part, and in this way, we may strengthen peace, guarantee prosperity and extend freedom to all those who share our aims."

Dr. Victor H. Reis, assistant to the director of national security and space at the White House Office of Scientific and Technological Policy, wrote: "The investment in a large base would have given NASA the possibility to play the role of a powerful political organization able to produce significant social and political benefits both domestically and internationally. This was the main reason the president said yes to starting the station."

STATION ARCHITECTURE

The White House directive gave the new program absolute priority and allowed construction of the first stable and secure manned settlement in space to start after decades of discussions and dreams. But the story that began the evening of January 25 under the white dome of Congress in Washington would be complicated and full of surprises.

At this point, work at NASA was proceeding smoothly, with adequate funding. Future potential users in the United States were consulted to find out their needs and specifications. A series of workshops held in the early months of 1984 defined the station's architecture: a collection of automated, interdependent, inhabitable orbiting structures in different orbits, with a manned base as the primary component.

The resulting concept was that of a station in an orbit inclined 28.5° to the equator and consisting of a service module, one or two laboratory modules, two logistics modules for Earth-space-Earth transport and a habitation module. It would be possible to attach further payloads to the exterior, and available power would be a maximum of 75 kilowatts. The crew would range from six to eight astronauts. In addition, there would be two automated orbiting platforms: one flying close to the mother station and another in polar orbit. The architecture of the system could grow in the future to increase the available power to 160 kilowatts, the number of astronauts to between 12 and 18, as well as the number of laboratory, logistics and manned modules. The platforms near the base would vary, and an Orbital Transfer

The configuration initially chosen for the station—the Power Tower. Manned modules are at the bottom.

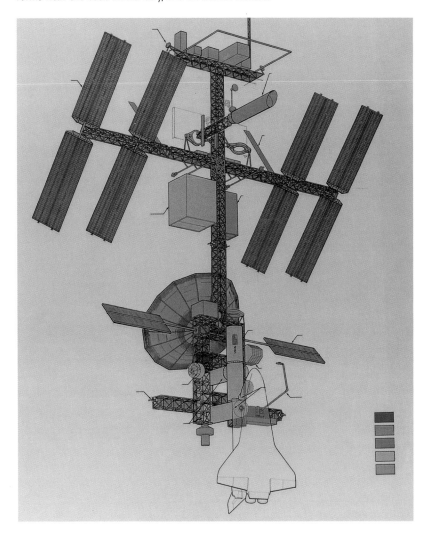

Illustration of the Power Tower, the configuration initially chosen for the space station.

Vehicle (OTV) would be available for transferring satellites and orbiting structures between different orbits. The use of robotic Orbital Maneuvering Vehicles (OMV) was also considered for service operations (like changing the instruments) on distant satellites and platforms. At the same time, support, inspection and maintenance activities would be carried out onboard.

Right from the start, it was established that supply flights carrying consumables (food, clothes, air, other gases and engine fuel) and necessary equipment should take place every 90 days. This would also offer the chance to change crews.

SIX POSSIBLE CONFIGURATIONS

Naturally, a station like this did not have a "unique solution" for its configuration, and in fact, the studies produced six designs, each with special characteristics that favored certain possibilities but obviously also had disadvantages.

1) Raft: a central body with pressurized and unpressurized sections to which were attached radiators and solar panels on partly (indeed, sometimes overly) flexible arms. The Raft was aerodynamically stable, and its position could be controlled both in inertial mode and

with respect to Earth. Operating in different attitudes widened the possibilities. The distribution of the external elements allowed the shuttle to dock safely and permitted the growth of the central core.

2) CDG-1: similar to the Raft but designed to operate only in a fixed position with respect to Earth. A downward-pointing trellis hung from the base of the central core, while other modules were placed at the end for stability.

3) Spinning Array: four empty shuttle tanks arranged in a cross and linked around a ring of solar cells. The whole assembly rotated to keep some of its solar cells always pointing toward the Sun. At the center of the wheel was a hub of manned modules that turned in the opposite direction so as to effectively remain still and yield the desired zero-gravity conditions (rotation, on the other hand, would give a level of artificial gravity). This design proved difficult to assemble.

4) Power Tower: designed to remain fixed with respect to Earth, like the CDG-1. However, its solar panels and radiators were located away from the manned core to minimize interference during shuttle visits. The modules were located at the base of the central structure for stability. All the elements were rigid, and position control was simple. Assembly required some in-orbit assistance from the shuttle.

5) and 6) Streamlined and Triangular: two rigid configurations that reduced problems of stabilization. The first consisted of a large, flat surface covered with solar cells on the top and radiators on the bottom. It operated in a fixed position with respect to Earth, with the large plane perpendicular to the orbit. The Triangular design had solar panels on the side facing the Sun and radiators on the opposite side. The station was always oriented toward the Sun, with its cells constantly illuminated.

THE POWER TOWER IS CHOSEN

After the president's announcement in 1984, NASA set up the Office of Space Station at its Washington headquarters to manage the vast undertaking. It was also established that the Johnson Space Center in Houston would direct the work, while major areas of responsibility would be given to the other centers. A large NASA team gathered in Houston to work on an invitation for proposals for the preliminary design and definition phase, to be distributed to corporations in September 1984. The station configuration would be chosen from the original six. This was then reduced to three—CDG-1, Triangular and Power Tower. Out of these, the Power Tower was the winner, and in the following spring, the first eight 18-month research contracts were awarded.

The Power Tower was 100 meters long, with a high central beam housing four pairs of solar panels, each having two surfaces 24 meters long and 10 meters wide. Six cylindrical modules 4 meters wide and of varying length were attached to the base of the vertical structure: two for habitation, two laboratories for research in life and materials sciences, one logistics module for supply missions and one for storage. Seven shuttle flights would be needed for complete assembly by 1992 at an altitude of 480 kilometers. However, after the fifth flight, the base would be permanently inhabitable by six astronauts. The total power supplied by the eight solar panels would be 75 kilowatts, and there was prospect for growth: up to 11 modules, 300 kilowatts of available power (thanks to replacing the solar panels with parabolic collectors) and 18 astronauts onboard.

Two hangars were planned to occupy the central part of the structure: one to store fuel and a smaller one for spare parts. Scientific instruments, especially for astronomical observation, could be installed on the smaller, transverse, upper beam.

THE INTERNATIONAL PARTNERS

Europe, Japan and Canada: Following the president's invitation to friendly nations to be involved in the construction of the station, negotiations began in 1984 with Europe (through the ESA), Japan and Canada to assess interest and ways of participating. A

Illustration of the Power Tower station design, showing the shuttle docked with one of the manned modules.

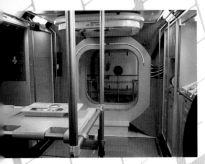

Background sketch: the Dual Keel station configuration, selected after abandonment of the Power Tower concept. Above, three details of the inside of a manned module, with the cupola through which the astronauts could look out.

first agreement was reached in the spring of 1985 and was laid down in a memorandum signed by the partners sharing out risks and benefits. After another year of study and negotiations, a multilateral intergovernmental agreement was reached in June 1988; and bilateral memoranda were finally signed in Washington on September 29, 1988. In any case, a presidential directive on the National Space Policy of the preceding February had reconfirmed support for the space station and the importance of relations with friendly countries.

The contribution of the international partners was valued at $8 billion, guaranteeing them a 28.6 percent share in use of the station—12.8 percent each for Europe and Japan and 3 percent for Canada. The United States held the remaining 71.4 percent. The same percentages were used to share out the common costs of the enterprise.

The various countries that had decided to join this enormous undertaking had differing interests. At the January 1985 meeting in Rome of the council of research ministers from countries in the European Space Agency (ESA), Europe elected to construct a laboratory module for the station, as well as an autonomous module orbiting near the main base and an automated orbiting platform. This would allow them to draw on the experience acquired with the Spacelab laboratory built by ESA for the shuttle. The decision was confirmed at the next ministerial conference in November 1987 in The Hague. Japan expressed a similar intention—its National Space Development Agency (NASDA) would also build a laboratory module. The Canadian Space Agency, on the other hand, would make a mobile robotic arm (Mobile Servicing System) for external work.

THE STATION CHANGES:
THE BIRTH OF "DUAL KEEL"
The year 1986 was a bad one for NASA and the United States. On January 28, the shuttle Challenger with its crew of seven astronauts exploded in the skies above Cape Canaveral 73 seconds after launch. This led to a general rethinking of America's commitment to space, and activities were paralyzed because the transport

into orbit of people and satellites was linked to the shuttle fleet.

However, the space-station program seemed to continue its progress, although a March review changed the planned configuration to ensure better microgravity conditions (that is, closer to zero gravity) inside the laboratory modules. The scientists had suggested this to improve the results of their experiments. The possibilities for external attachments were also broadened, and a more efficient location was found for the maintenance hangar.

In short, the design was revolutionized. The Power Tower was abandoned, and a new Dual Keel configuration was born. This consisted of a horizontal trellis structure with two pairs of 74-meter-long solar panels at each end to provide 75 kilowatts of power. The nucleus of laboratories with lateral radiators would be located at the center of the 108-meter-long horizontal structure.

Given the planned presence of European and Japanese laboratory modules, NASA reduced its own to only one. To this were added a habitation module, a logistics module for transportation and four "nodes"— manned mini-modules for the attachment of the larger ones. The planned crew size was eight astronauts.

The station, now christened "Freedom," could grow thereafter with the addition of two vertical structures (hence the name "dual keel") 105 meters high. The structures would be closed off above and below by two 45-meter elements to form a rectangle around which hangars, scientific equipment, systems for the docking of automatic vehicles and antennas could be located.

The modified design specifications were concluded by the end of January 1987, permitting a reliable estimation of costs and a precise definition of the technical aspects. These were vital prerequisites for the awarding of development contracts in December 1987 to the companies involved. The National Research Council (NRC), an important scientific body, also approved the modifications in that same year.

SPACE STATION FREEDOM: WORK ORGANIZATION
NASA had set up a three-tier hierarchy of jobs and responsibilities. The first tier was program direction,

under Washington headquarters. The second level concerned the administrative and technical management of the program and was assigned to the Space Station Freedom Program Office in Reston, Virginia. The third and final tier was management of the program's definition, development and operations phases, and responsibilities were distributed among the various NASA centers that supervised the activities of the companies building the different elements. (The management scheme adopted by NASA was the result of a precise recommendation by the presidential commission on the Challenger accident.)

The Marshall center in Huntsville, Alabama, was responsible for the habitation and laboratory modules, the nodes that connected them, the logistics systems and life-support systems. This was gathered under the heading "Work Package 1," with construction by Boeing and its subcontractors.

Work Package 2 was the responsibility of the Johnson center in Houston, Texas, and covered the beam structure, propulsion systems, data management, temperature control, communications, navigation and guidance systems, as well as extravehicular activity.

The corporate partner in this was McDonnell Douglas.

Goddard Space Flight Center (with corporate partner GE Astro-Space) was assigned Work Package 3, which covered service structures, robotic devices, payloads and orbiting platforms.

The Lewis Research Center in Cleveland, Ohio, was responsible for power generation, handling and distribution (Work Package 4), and the equipment would be built by Rocketdyne. The Johnson center (with the support of Rockwell) was further responsible for mission management, ground control and astronaut training, while the Kennedy Space Center in Florida had to manage payload preparation and launch operations, aided by McDonnell Douglas and Harris.

SERIOUS CRISES AND DESIGN CUTBACKS

Beginning in 1987, the station program entered a critical phase that showed the management structure to be complex, costly and de facto unproductive. In fact, in 1990, during its discussion of the budget for the next fiscal year, Congress ordered a redesign with the aim of reducing development costs. It went so far as to ask for cuts of $6 billion over the next five years.

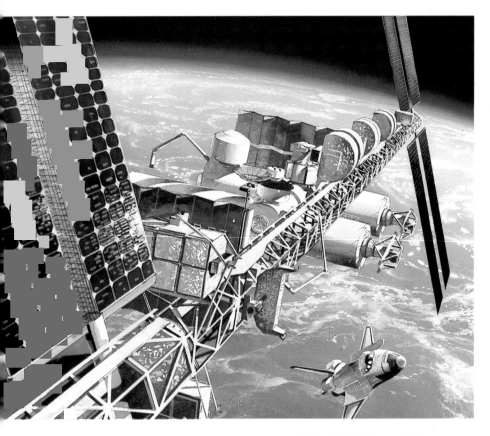

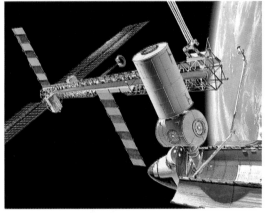

Three illustrations of space station Freedom. The one in the middle shows the base during assembly.

From the initial projection of $8 billion, the cost of the station had rocketed to $14.5 billion, and after six years, it was still only a "paper station."

In March 1991, NASA (with the other participating space agencies) completed the restructuring, incorporating changes that made Freedom more economical, smaller and easier to assemble in orbit. The pairs of solar panels were reduced to three, decreasing the power supply to 56 kilowatts. The number of astronauts to be accommodated onboard was cut by half (to four), as was their planned extravehicular activity, because this had caused a lot of argument and a string of criticisms about its high cost. The station's mission was also revised, now basically limited to research in life and microgravity sciences relating to the production of materials and substances in conditions of near weightlessness. Technological experiments and trials for human survival in space could be added later. This was far from the rich, eight-point mission model approved in 1984. Now there were only two points.

Europe was also affected by the crisis, and similar revisions were made. One of the two manned modules that ESA had meant to add (the "free-flyer," which was to have flown at a distance from the station) was canceled. The European "attached module" Columbus that was to be berthed to the station was also reduced. Japan, on the other hand, did not have the West's difficulties and continued its work undisturbed.

Yet despite the revisions, space station Freedom was actually on the way out. The year 1991 saw the United States involved in the Gulf War. President Bush entered a period of political weakness, and consensus was increasingly uncertain. In 1992, he failed to be reelected, and the White House was occupied by Democrat William ("Bill") Clinton. This, together with the collapse of the USSR, radically affected the future of the American space-station program, and its probable cancellation was being discussed.

Cutaway view of Freedom's manned module.

The Soviet station Mir 2 and the Buran shuttle

In the 1980s, the Soviets designed a new station, Mir 2, illustrated far right, which was to continue the work of Mir 1. In the meantime, the reusable shuttle Buran (Snowstorm) was being built. It only made one orbital flight, without a pilot, on November 15, 1988; it is shown before launch in the photograph at right. This first and only flight lasted 205 minutes, with takeoff and landing at Baikonur. Buran weighed 100 tonnes and was launched on the giant Energia rocket.

The various components of

Mir 2 were to have been carried in Buran's cargo bay and then assembled in space. Cosmonauts had already been selected and were being trained to pilot the

Soviet shuttle, which was very similar to its U.S. counterpart.

Buran was 37 meters long, and its widest diameter was 23 meters.

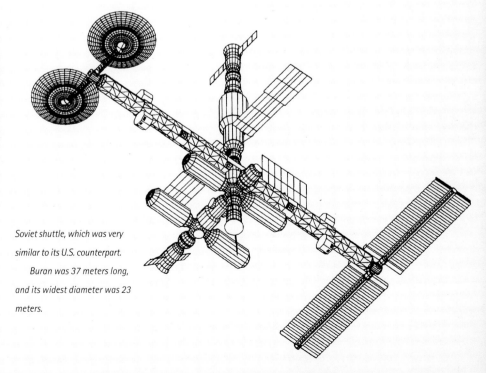

6

Cooperation in space: ISS

The International
Space Station

Cooperation in space: ISS

The year 1992 brought dramatic changes. Russia, now a republic, began a new era under Boris Yeltsin, who had succeeded Gorbachev. The space agency RKA was created in Moscow to manage all civilian space activities, with Yuri Koptev appointed as director.

The Gulf War was over, and U.S. President Bush was preparing for new elections. In March, he approved the Space Exploration Initiative (SEI), whose goals included a return to the Moon and the exploration of Mars. But the initiative seemed to be limited more to giving a direction to the plans rather than to realizing them. Moreover, the estimated costs—hundreds of billions of dollars—were prohibitive and therefore unpresentable to Congress. This, plus Bush's electoral defeat and exit from the White House, led to the cancellation of the SEI.

The beginning of 1993 was a time of critical decisions for the American and Russian space programs, and it proceeded to become a key year for the birth of a large orbital base. On March 9, the new American president, Bill Clinton, asked NASA (now headed by Daniel Goldin) for a 90-day review of the Freedom space-station program, with a view to reducing costs and construction times. Goldin therefore created a Station Redesign Team within NASA made up of 45 technicians and 10 experts drawn from the international partners. Meanwhile, on March 25, the American vice president set up a committee (Advisory Committee on the Redesign of Space Station) chaired by Charles Vest, president of the Massachusetts Institute of Technology, and comprising 16 representatives from various academic, corporate and military organizations to supervise the work of the NASA team.

Preceding pages: artist's impression of the ISS.

The symbol of the International Space Station. The background stars represent the objectives of the conquest of space. The chain bordering the image of Earth symbolizes the millions of people around the world who have worked to build the station. The five-pointed stars represent the five space agencies (NASA, RKA, ESA, NASDA, CSA) that have participated in the project. Finally, the laurel leaves symbolize mutual peace.

THE SELECTION OF ALPHA

Keeping to the deadline, the technical group presented its proposals to the Vest committee on June 7, and three days later, the committee submitted its final report to President Clinton for a decision. The team and the committee had drawn up three plans, called simply A, B and C. The committee recommended adoption of plan A, or Alpha, because it was a modular concept that would best apply the work already done, it would reduce costs and it respected agreements with the international partners. The recommendation was put forward by the president himself, and he made the choice official by ordering drafting of a detailed working plan within 90 days.

To do their job undisturbed, the experts gathered at Crystal City, an office complex on the outskirts of Washington close to the National Airport. Meanwhile, a series of political maneuvers between Moscow and Washington attempted to forge a collaboration between the two countries. The White House had stipulated this as a fundamental objective on which the real rebirth of the station program depended.

A meeting between Bush and Gorbachev in 1991 had tried to breathe new life into the agreement on space cooperation signed in 1987 that had reestablished space relations between their two countries after a period of total blackout. The new agreement, which survived the collapse of the USSR, anticipated international crew exchanges (Americans on Mir and Russians on the shuttle). It was ratified in June 1992, to be extended over five years. So NASA and the RKA began a new era of cooperation that saw cosmonauts Sergei Krikalyev and Vladimir Titov preparing for flights on the shuttle. At the same time, plans were made for a 90-day stay by an American astronaut on Mir, and an orbital rendezvous between the shuttle and the Russian station was proposed for mid-1995.

But in a certain sense, these were only the trimmings. Washington and Moscow were in fact considering something more substantial that was to have significant consequences. What was being investigated was how to incorporate Russian technology and know-how into the plans for the new Alpha station. It started with the simple acquisition of Soyuz capsules for use as emergency rescue vehicles; within a short time, however, there developed an understanding that the two countries would build a common orbital base together.

An important factor in the negotiations was that the Russians were already working on a new station, Mir 2. But given the critical economic situation after the dissolution of the USSR, they would never be able to finish it alone.

Everything was expedited by a new agreement in July 1993 that extended the one of the previous year. Consequently, a Russian Integration Team joined the Americans and their partners at Crystal City, and together, they fleshed out what in previous weeks had only been an idea. On August 30, a new joint plan was ready.

RUSSIA JOINS THE PROJECT

The opportunity to make the plan public came three days later on September 2, during the first meeting of the joint Russian-American commission on economic and technological cooperation. Chaired by Vice President Al Gore and Russian Premier Viktor Chernomyrdin, the meeting concluded with the signing of numerous understandings in various areas, including space and the construction of the first "cooperative house" in orbit. The signing was historic in that it marked the end of an era of antagonism and the beginning of a new phase of cooperation in a field whose technological, political and military nature had hitherto made it

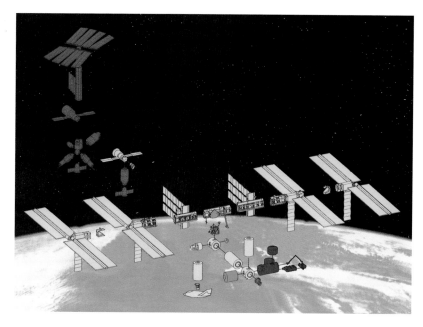

an arena for confrontation and dispute. Until now, space programs had been viewed as an ideal means of demonstrating the superiority of a political system. Now they became a proving ground for experiments in cooperative agreements that could be usefully applied to other fields.

Above, the elements making up the International Space Station. Below, sketch of the completely assembled base.

Main specifications:

length: 108.4 meters

width: 74 meters

mass: 419 tonnes

orbital altitude: 370-460 kilometers

orbital inclination: 51.6 degrees

pressurized volume: 1,200 cubic meters

crew: 6 astronauts

available power: 78 kilowatts (maximum 110 kilowatts)

The Gore-Chernomyrdin plan had three phases. The first included two years of American astronaut missions on Russia's Mir station, joint use of the Mir modules Spektr and Priroda (still on the ground at that time) and the development of new technologies. This phase would be sustained by funding from both NASA and the RKA to the tune of $100 million a year for four years.

The second and third phases would be dedicated to building the new International Space Station. The plan was defined in a document signed on November 1 by representatives of the two space agencies. In the meantime, work proceeded on the international plan to include Russia's presence in a project which up to then had been managed by the United States and shared by Europe, Japan and Canada.

The three non-U.S. partners did not exactly welcome either the decision to redesign the station yet again (announced by Clinton in February 1993) or the subsequent entry of Russia, which would inevitably relegate Europe, Japan and Canada to third place. In any case, the two major players were bound to be America and Russia for political reasons; furthermore, the Russians had experience with space stations and lengthy manned missions that the United States did not. But the three other partners had guaranteed their participation in the Freedom program to the tune of $8 billion, of which $3.2 billion had already been spent. Therefore, changing plans and roles did not please them.

CRISIS AND RECOVERY
AMONG THE OTHER PARTNERS
To find a solution, the American State Department organized two consultations with the partners in Washington in May and June, followed by a third intergovernmental meeting in Paris in October, where the partners declared their willingness to look into the idea of allowing Russia into the project.

It was the Europeans who were having the most difficulty at the time, while the Japanese proceeded without any great problem. NASA maintained that the loss of the Old World countries' collaboration would have a negative effect on the entire project and on the

Stations in science fiction: *Star Wars*

The first film in George Lucas's Star Wars series came out in the United States in 1977. The Star Wars trilogy was such a success that it has recently been revived and continued. The story: a group of rebels led by Princess Leia Organa fight against the all-powerful Galactic Empire. Their objective is to destroy the Death Star station. The heroes are young Luke Skywalker, the old Jedi knight Obi-Wan Kenobi, adventurer Han Solo and droids R2-D2 and C-3PO.

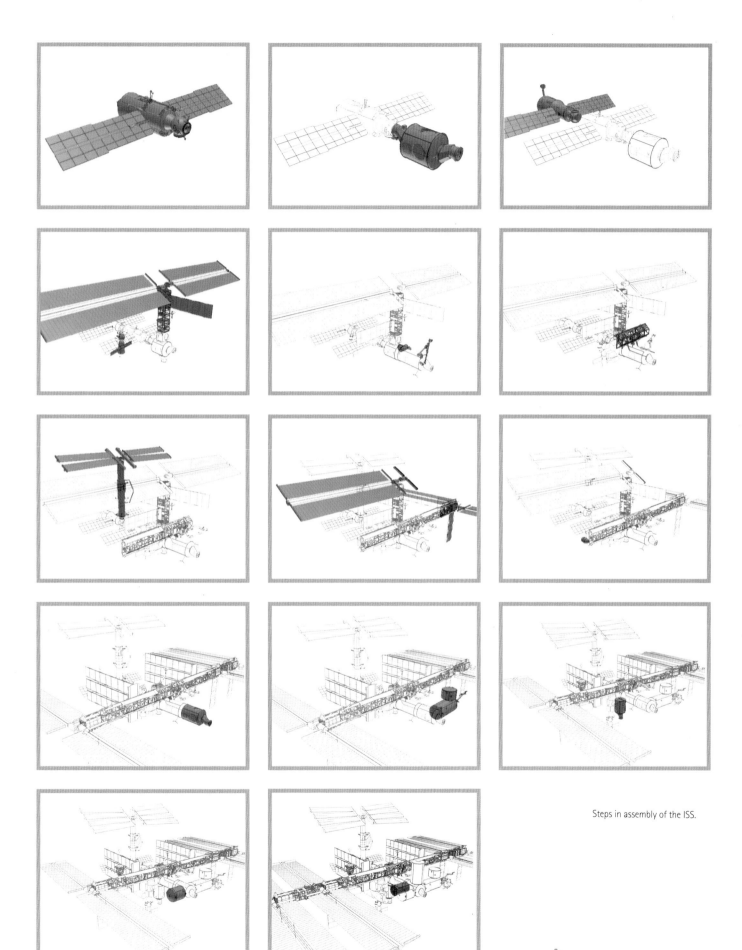

Steps in assembly of the ISS.

other participants. The truth was, however, that Europe could not think of leaving the project, especially after the entry of Russia, if it wished to maintain an international role in space. The alternative was withdrawal from manned flights, because Europe was unable to build an autonomous station. Even attempts at bilateral solutions, as the Germans tried briefly with the

Russians, proved ineffective.

Moreover, in the early 1990s, the various European countries had fallen into a space-program crisis, primarily financial in nature but also regarding content and strategies, that was forcing a serious review of the station program. The original plan had incorporated a European-made module attached to the American station, and a second, free-flying module that would be visited from time to time—in effect, an autonomous mini-station. A planned polar platform was also later removed from the station program and built separately. In addition, the design phase for the Hermes reusable mini-shuttle had gotten under way but was halted after a conference of research ministers in Granada in November 1992 because of escalating development costs. Just as NASA had reshaped its ambitions, so ESA first canceled the free-flying module and then reduced the planned size of the attached module, the Columbus Orbital Facility (COF).

RUSSIA IS IN

In the end, however, on December 6, 1993, Europe, Japan and Canada issued a formal invitation to the Russian government to join the group of countries committed to the undertaking. Immediately afterwards, at the end of the month, the Johnson center in Houston began revising the project, taking into account all the protagonists, old and new. The plan presented in September, after 90 days of drafting, went beyond plan A to include some aspects of plan B that would better accommodate the diverse needs of the countries involved.

A NEW PARTICIPANT

At the beginning of 1994, while the engineers worked on detailed design plans for the new International Space Station, the political negotiations continued to define reciprocal roles, rights and responsibilities.

Canada and Japan. In 1994, Canada declared its intention to reconsider its participation as builder of a robotic arm. But the Canadians' difficulties were overcome when NASA made some concessions in the execution phase, and work continued without further problem.

As noted earlier, Japan displayed no uncertainty

Right, the FGB/Zarya module.

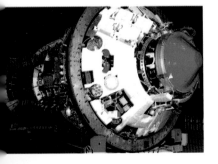

Above, the two ends of the module.
Right, integrating the module in Moscow.

and proceeded with the work agreed upon, indeed preparing to increase its commitment. Japan was already providing a module called JEM with an attached logistics module, an external platform and a robotic arm. In 1997, the Japanese space agency NASDA reached an agreement with NASA that it would also build the centrifuge module (initially to have been an American product). In effect, the deal would compensate the Americans for the cost of launching the JEM on the shuttle. Meanwhile, NASDA began making plans for a pressurized vehicle (to be launched with the Japanese H-2 rocket) to take equipment and other materials to the station.

Europe. Europe, on the other hand, had to wait until October 1995 for its contribution to the station. After two years of internal discussions, the council of research ministers overseeing ESA activities met in Toulouse and definitively approved their participation plan.

The budgeted expenditure would be 2,651 million ECU (in 1995 values), to which was added 207 million ECU for microgravity research to be conducted on Europe's Columbus Orbital Facility (COF) module. The three countries with the highest commitments were Germany (41 percent), France (27.6 percent) and Italy (18.9 percent). In addition to the Columbus module,

The Proton rocket used to take FGB/Zarya into orbit from Baikonur.

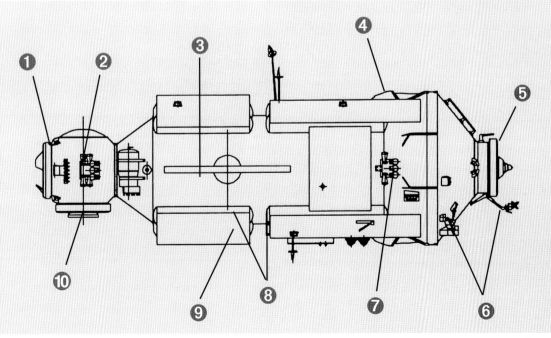

Sketch of the various elements of the FGB/Zarya module:
1) forward docking system
2-4-9) rocket engines for position control
3) solar panel
5) aft docking system
6) passive docking system
7) radiators
8) fuel tanks
10) antennas for the Kurs electronic docking system
Main dimensions:
length: 12.3 meters
diameter: 4 meters
mass: 19.5 tonnes (in orbit)

turn made the Alenia Aerospazio company responsible for it.

Earlier, with the same notion of bartering, the European agency had provided NASA with an experimental apparatus called Glovebox, a mobile support system for external instruments (Exapod) and a –80°C freezer, all to enable partial use of the station before the launching of the COF.

Russia. Russia's participation in the project would be considerable. Besides the first station module—the FGB, built in Russia with American funds—there would be a service module, a module for multiple attachments, a scientific platform with solar-power panels, at least two science modules, the Progress automatic resupply vehicles and manned Soyuz capsules for transport and rescue craft, at least in the beginning.

Node 1 (Unity) under construction, above, and being prepared at Cape Canaveral, right. Standing in front are members of the crew of the shuttle Endeavour (STS-88) that would take it into orbit (from left to right): Frederick W. Strucknow (pilot), Nancy Jane Currie (mission specialist), Robert D. Cabana (commander) and James H. Newman (mission specialist). Bottom right, drawing of Unity.

Main dimensions:

length: 6.6 meters

diameter: 4.2 meters

ESA was to build an Automated Transfer Vehicle (ATV), which would be used for both the transport of materials to the station and to boost its orbit as needed. ESA was also to supply the data-management system for the Russian service module and a robotic arm attached to the Russian power tower.

Later, NASA and the European agency made a deal that solved the problem of the cost of launching the Columbus module with the American shuttle. Instead of paying money, ESA would supply NASA with two station nodes (numbers 2 and 3), which would otherwise have had to be built in the United States, together with onboard equipment like refrigerators and freezers. The construction of the nodes was later assigned by ESA to ASI, the Italian space agency, which in

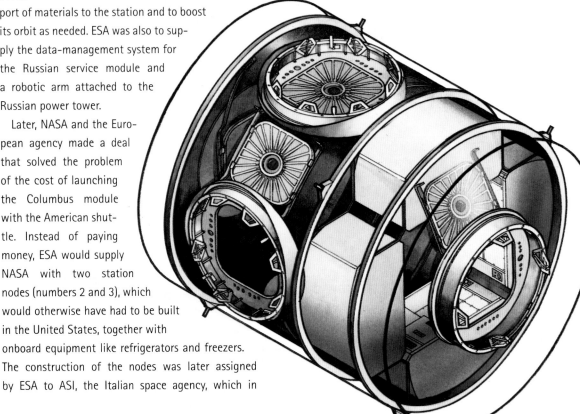

Italy. In 1991, the Italian space agency ASI signed an agreement with NASA to develop and construct two pressurized logistics modules that would be carried by the shuttle to transport materials and instruments to and from the station. Once built, they would be consigned to NASA in exchange for station use time and flight opportunities for Italian astronauts, who were now part of the ESA astronaut corps. The agreement was revised in 1992, increasing the number of modules to three and adding the stipulation that Italy would create a technical center in Turin to support the orbital missions and run the Italian experiments on the station.

Italy was also involved in a European context through its participation in the Columbus module, to which was later added construction of two of the nodes. In summary, four manned structures for the station would be Italian-made (six if we count the logistics modules). This was a result of Italian expertise in constructing manned cylindrical structures— namely, ESA's Spacelab module and Spacehab, a

module built for the American Spacehab company, both of which had been used in the shuttle.

Brazil. In July 1997, Brazil joined the project with a letter of intent by which it would supply parts of the station in return for use time and flights by some Brazilian astronauts. Brazil would provide the Technological Experiment Facility (TEF) and an attachment system for the station's Express Pallet. In addition, they would provide an instrument for Earth observation, and work was begun on an unpressurized shuttle

Above left, signing of the agreement for construction of the ISS at the State Department in Washington on January 29, 1998 (from left to right): Yuri Koptev, director of the Russian RKA; Antonio Rodotà, director general of Europe's ESA; Daniel Goldin, NASA administrator; Willam Mac Evans, president of Canada's CSA. Japan would sign later.

Above, document commemorating the event signed by Daniel Goldin.

Far left, in December 1998, the crew of STS-88 began assembly of the ISS, joining the U.S.-built node Unity to the Russian-built module Zarya. James Newman works on Unity's communication cables as crewmate Ross looks on.

Left and below, Node 1 (Unity) being prepared before launch.

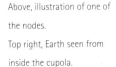

Above, illustration of one of the nodes.

Top right, Earth seen from inside the cupola.

transport module—Unpressurized Logistics Carrier (ULC)—to transport equipment that did not require a pressurized environment.

THE FINAL SIGNING

At last in 1998, on the eve of the start of in-orbit assembly operations, all the players in the undertaking met in Washington to sign the last act defining roles, responsibilities and reciprocal rights and duties "according to international law." This replaced the document signed 10 years earlier, on September 29, 1988, because in the interim, both the station and its participants had changed.

On January 29 in the Dean Acheson Auditorium at the State Department, in the presence of Vice Secretary of State Strobe Talbott, the research ministers and government representatives of the 15 participating nations (the United States, Russia, Japan, Canada and the ESA member states Belgium, Denmark, France, Germany, Italy, the Netherlands, Norway, Spain, Sweden, Switzerland and the United Kingdom) signed the

Agreement for Intergovernmental Cooperation on the Space Station.

Immediately after, the heads of the respective space agencies signed three bilateral memoranda to govern their work. NASA was headed by Daniel Goldin, the RKA by Yuri Koptev, the ESA by Antonio Rodotà and the Canadian Space Agency (CSA) by Mac Evans. Japan's NASDA would sign the documents at a later stage. This agreement was historic in that it set the stage for launching the greatest work of human engineering of all time and was the broadest international-cooperation agreement ever achieved.

STATION IDENTIKIT

Construction costs and time lines. The new Alpha station approved by the American president in 1993 was based on a strict budget. The project had to take into account that $10.2 billion had already been spent on various studies, designs and tests, and attempts should be made to apply what had already been done. Alpha did in fact use 75 percent of the work done for Freedom. Moreover, time lines had to be strict. In nine years, starting in 1994, the station must be designed, constructed and assembled in orbit, with completion by 2002.

Overall spending was fixed at $17.4 billion—$6.4 billion for development and construction and $11 billion for orbital transport and assembly. A further $13 billion had to be added for the expected costs of running the completed station for 10 years, from 2003 to 2012.

Investments by the three "veteran" non-U.S. part-

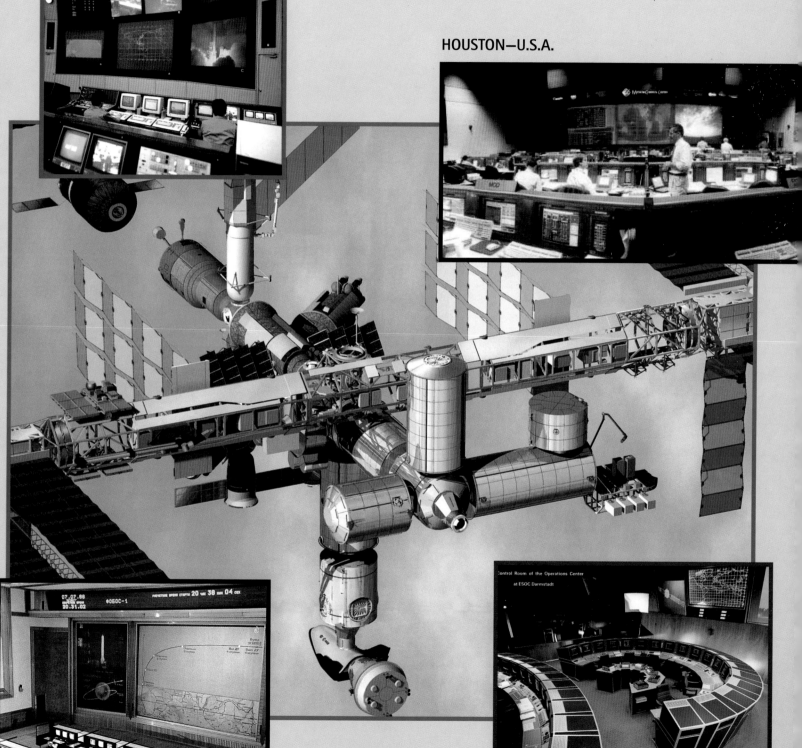

TSUKUBA—JAPAN

Control centers for the
International Space Station.

HOUSTON—U.S.A.

KOROLEV—RUSSIA

DARMSTADT—GERMANY

Right, the Russian service
module under construction
in Moscow.

Main dimensions:

length: 12.9 meters

diameter: 4 meters

mass: 19 tonnes (in orbit)

solar-panel span: 29.25 meters

Below, illustration of the first
three elements of the station
(from left to right): the
Russian service module, the
FGB/Zarya module and the
American Node 1 (Unity).
Docked below is a Russian
Soyuz capsule.

ners were estimated at around $8 billion, although it was difficult to evaluate the Russian contribution. Although $13.4 billion had been spent up to 1993 ($10.2 billion by NASA and $3.2 billion by ESA), the total cost of the station (excluding the Russian contribution) up to its completion in 2002 was expected to amount to $30 billion. Construction was expected to take 18 years from President Reagan's launching of the initiative. According to NASA calculations, Russia's participation would mean a saving of around $2 billion.

In March 1998, NASA heads confirmed during a hearing in Congress that there had been an increase in project costs by almost $4 billion, making the expected total $21.3 billion instead of the originally planned $17.4 billion. The new figures came from an investigation made by an independent committee under Jay Chabrow. Factors included new work on a rescue vehicle, a rise in construction costs, delays in Russia's launch of the service module and development of an automatic resupply vehicle that previously

Below, drawing of the
American laboratory module.
Top left, a crew member
aboard Discovery (STS-96)
recorded this image of the ISS
on June 3, 1999, with a 70mm
camera following separation
of the two spacecraft. Lake
Hulun Nur, in the People's
Republic of China, is visible in
the lower left portion of the
frame. Work performed during
the May 30 space walk by
astronauts Tamara E. Jernigan
and Daniel T. Barry, including
the installation of the Russian-
built crane (called Strela)
and the U.S.-built crane, is
evident at various points on
the ISS.
Main dimensions:
length: 9.9 meters
diameter: 5.2 meters

had not been considered. This also increased the expected in-orbit assembly time by one year.

WHO LEADS THE ENTERPRISE?

The new project required reorganization and simplification of how the work was to be managed. No longer would various NASA centers be working at the same level; instead, only one—the Johnson center in Houston—would be the leader, while the others would have more limited responsibilities. This decision, announced in August 1993, was extended to the corporate participants so that the various major contractors, including the foreign firms, would be directed by only one—Boeing.

In 1995, NASA awarded Boeing a contract worth $5.63 billion, later increased to $7 billion. In 1997, 5,000 Boeing workers were directly employed on the station. Added to these were another 5,000 employees of the other American subcontractors, plus about 30,000 workers who were involved indirectly. On top of that, there were all the people working in Japan, Europe, Canada and Russia. It is obvious, therefore, that management of the project was a difficult challenge.

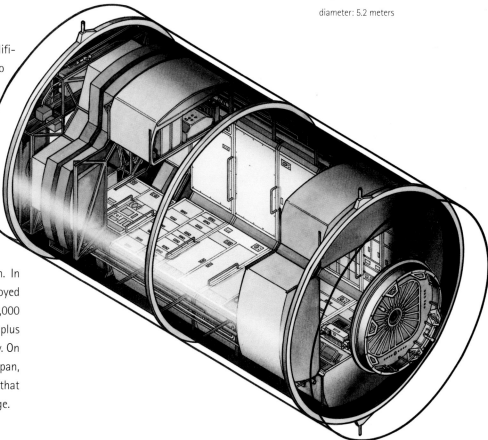

Top to bottom: two pictures of the interior of the American laboratory; construction of the U.S. laboratory; the Express Rack for small, short-term experiments.

THE DESIGN

With the entry of the Russians, the working name Alpha was dropped in favor of the more generic International Space Station. It would draw upon all the previous experience the various players had acquired in the field of life in orbit, especially the Russians with Salyut and Mir and the Americans with Skylab.

The discussions at Crystal City, starting with the Alpha project, had resulted in a base that incorporated components of Mir 2, which the Russians had already begun to build (with plans to assemble it in orbit using their Buran shuttle). The central component of the station was to be a beam with two sets of solar panels at the ends. A cluster of manned modules and connecting nodes would be anchored across the middle of the beam, with docking systems for American shuttles and Russian Soyuz and Progress vehicles.

The station concept was rather like a gigantic Lego construction, and at least 50 American and Russian launches would be required to carry all the pieces to be assembled. Once completed, the giant piece of cosmic architecture would be composed of six laboratory modules: two Russian, two American (one with a centrifuge built for NASA by the Japanese), one European (COF) and one Japanese (JEM). There would also be an American habitation module and two Russian modules containing life-support systems.

The Russians, moreover, would provide the FGB module (the first piece to be launched) and a manned service module. Then there was the logistics module to be built by Italy for NASA and reserved for the transport of materials. Various mini-modules, or nodes, would serve as connecting and habitation segments between the other modules.

Now, let us look at the birth of this cosmic castle and its characteristics.

THE SECOND PHASE: CONSTRUCTION

Whereas the first phase of the project involved joint missions to the aging Russian orbital base Mir to learn how to work together and to get to know each other's technologies, the second phase involves actual in-orbit construction of the new base. This requires 10 launches to establish sufficient resources to begin research activities and have a stable living space. The base is to have four habitable elements: the Russian FGB and service module and the American Node and laboratory module. The beam, with the first solar-power panel, is also American.

THE FIRST MODULE: FGB/ZARYA

The first module to be launched (using a Russian Proton rocket) was the Functional Cargo Block (Russian initials *FGB*), now christened *Zarya* (Dawn). Built by the Khrunichev Center with American financing to the tune of $190 million, Zarya is the successor to the military modules studied by Chelomei in the 1970s and later used as laboratories on Mir. It has guidance systems, sensors and motors and is able to control its position during the first operational phases as well as supply some power via its two solar panels. Of course, it also has the necessary human life-support systems, although its primary function is to serve as a storehouse.

Zarya's equipment provides the environmental conditions required by the American Node, the second station element to be integrated. Zarya has a compartment to which external modules can be attached, and the Node likewise offers the possibility to attach other modules.

NODE 1: UNITY

Node 1, like the other two planned nodes, is particularly important because it will serve as a conduit for all the essentials (power, communication, data, liquids and gases for the environmental systems, etc.) needed by the other modules that will be attached to it. For this reason, its aluminum walls have six openings (two at the ends and four on the sides) with attachment systems. Therefore, it had to be constructed with enough strength to bear the stresses of loads placed in various positions. To give an idea of its complexity, it is made up of 50,000 mechanical parts and has 216 conduits for gases and liquids. Its 121 electrical cables placed end to end would stretch 4 kilometers, and 200 specialists are necessary to put all these pieces together correctly.

As well as all the electronic equipment, the Unity node also has two wardrobes for the American astronauts' clothes. The aft attachment system has a 2-meter-long asymmetrical conduit, the Pressurized Mating Adapter (PMA), which acts as a link with the Russian FGB/Zarya module. Inside, there are computers and electrical systems.

THE SERVICE MODULE

The service module, practically the same as the one on Mir, attaches to the end of the FGB. While the FGB is principally a storage module with fuel reserves and the capability to maintain correct position, the service module was designed as a habitation unit for the astronauts, where they would live, eat and sleep until the arrival of the American habitation module. Therefore, the service module contains two autonomous cabins for the astronauts, a lavatory, a shower and exercise equipment (a treadmill and a cycle). A table can be let down from the wall for meals, and there is a fridge to store food.

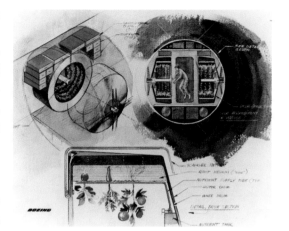

Two sketches showing the centrifuge module and its interior during the design phase.

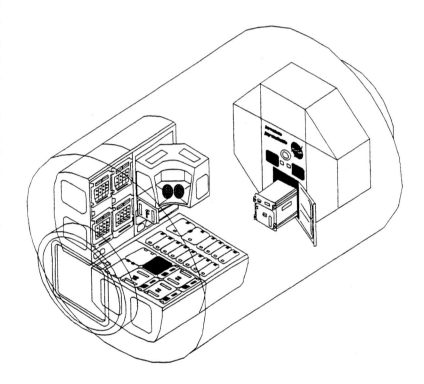

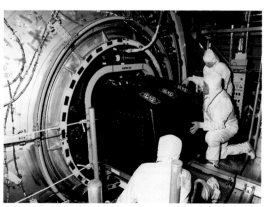

Left, installing the racks inside the American laboratory module.

All this equipment is nearly identical to Mir's. Some improvements have been made, but the environment is substantially unchanged in size (4 meters maximum diameter) and fittings, including the traditional control panel with its monitors at the forward end. The astronauts have 14 circular portholes to look out of, three of which (22.5 centimeters in diameter) are in the forward compartment to allow visual control during attachment maneuvers.

The service module was designed to take control of the station during the entire second phase, because it has the necessary systems: directional sensors, computer to calculate position, instruments and motors to adjust attitude. The station's "brain"—the entire onboard computer system for guidance, navigation and control of the station up to the arrival of the American elements (until that moment, control was from the FGB module)—was made by ESA and supplied to the Russians.

Because of Russia's delays in completing the service module, NASA developed an automatic module, the Interim Control Module (ICM), previously tested by the Naval Research Laboratory in Washington on naval observation satellites. This module, which cost about $100 million, was conceived as a temporary reserve for the Russian service module and could be attached to the FGB to boost the assembly's orbit.

THE AMERICAN LABORATORY MODULE

The fourth and final important element in the first phase is the American laboratory module. Launched by the shuttle, it has 5 of the 11 planned equipment units, or racks, for supplying power, controlling temperature and humidity and renewing the air. Another 13 racks will be used for scientific activities. The 19 racks needed to complete the system will be launched separately.

The module is composed of three cylindrical sections joined together with machined aluminum walls and internal systems to allow the astronauts to live and work in a shirtsleeve environment as in the Russian modules. The system consists of an oxygen distributor, a carbon dioxide collector that discharges externally, a temperature- and humidity-control system and a system for transforming excess humidity from the air into usable water. There is also a system

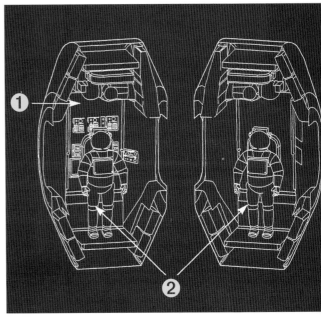

Left, one of the station's
solar panels completely
deployed during ground tests.

Below, docking system for the
modules during ground tests.

Left, airlock during construction.
Far left, cutaway view: 1) control panels, 2) space suits for
space walks.

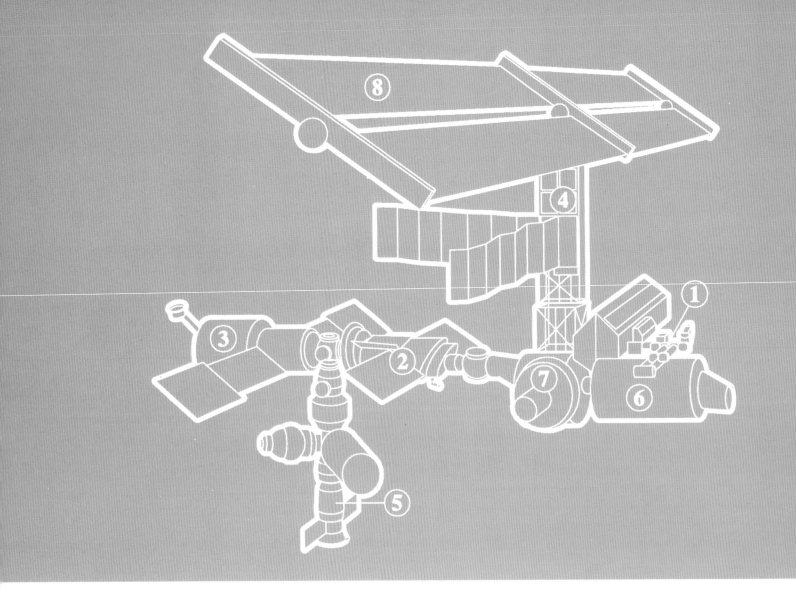

that can identify and remove any toxic atmospheric substances. A piece of equipment is to be installed at a later date to produce water with the carbon dioxide instead of ejecting it into space.

Finally, another system controls the internal pressure, which must remain the same as at sea level. In the event of a loss of pressure, the system automatically compensates by using nitrogen from the external tanks in the airlock area, where the emergency oxygen tanks are also stored. During the second phase, however, air control and exchange is done by the Russian modules, while communications are via Russian ground links.

The four walls of the module have movable, replaceable racks for a variety of experiments. The free space in the center, where the astronauts float about in zero gravity, is very square and relatively empty. A 50-centimeter porthole has been placed in the wall facing Earth for surface observations.

American scientists have prepared six main instru-

ments for the laboratory: one for biotechnology research, one to study fluids and combustion, a system of ovens (one supplied by ESA), equipment for biological experiments with plants and animals, a centrifuge and instruments to study the adaptation of the human body to zero-gravity conditions. This equipment occupies one or more racks as needed. The racks are of standard size throughout the station (except for the Russian portion) so that they can be used wherever they are needed. Finally, the module contains ESA's MSG rack. This allows astronauts to conduct experiments in microgravity by moving specimens with their hands placed in gloves attached to the MSG.

POWER TOWER AND GUIDANCE SYSTEMS

To supply electric power, there is a trellised tower, at the end of which are two sets of solar panels able to generate 20 kilowatts of power. This will later be removed and placed at the end of the truss (the station's skeleton) to become one of four pairs of solar panels.

A compartment at the base of the tower, installed on the Node before the tower, contains four Gyrodynes, large gyroscopes powered by an electric motor that orient the station. These are not activated until the arrival of the laboratory carrying their hardware and software.

In this phase, the base is guided by the Russian service module, whose four engines also serve to boost the orbit periodically as needed. The same operation can be done with Progress automatic vehicles, as was done for Mir.

THE CANADIAN ROBOTIC ARM

The main portion of the Canadian robotic arm (Mobile Servicing System) required for the assembly operations is also placed in orbit during the second phase. This is a complex robotic structure with three elements. One acts as the base, moving up or down to where it is needed on the long truss that makes up the skeleton of the station. The second part is the arm itself (brought into orbit in this second phase), divided into two 17-meter segments. This was derived from

the similar Canadarm built by the Canadian Space Agency for NASA's space shuttle. The third part is a sort of 3.5-meter-long robotic "hand" at the end of the arm. It has two movable fingers for precision work and manipulating. Sophisticated visual systems allow astronauts inside the station to monitor the arm's operations in detail. The arm can shift loads of 100 tonnes, while the "hand" can handle 600 kilograms.

Illustrations of the station after completion of the first assembly phase.

Above, detail of the robotic arm built by ESA and installed on the Russian Science Power Platform.

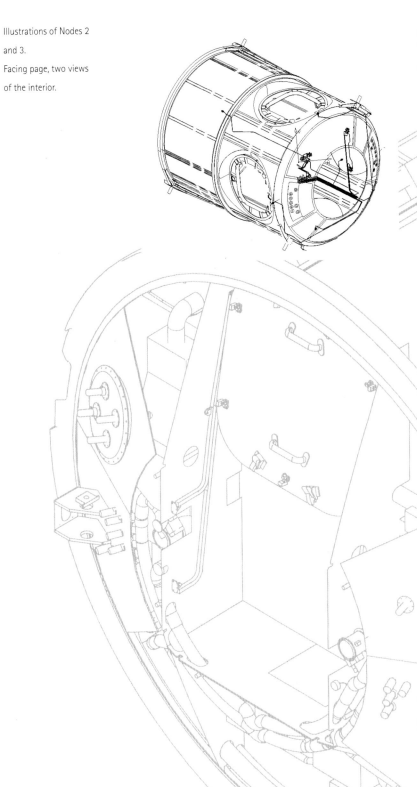

AIRLOCK

The last element to be added during the second phase of assembly—during which a crew of three astronauts/cosmonauts can live onboard—is an airlock compartment for access to the outside. The airlock is connected to one of the side exits of the Node, and around it are the oxygen and nitrogen tanks needed for life support. At this point, the International Space Station will have reached the minimum size to allow the long-term presence of humans. Although there will be a limited crew at this stage, it will be sufficient to start the planned experiments and research.

LEONARDO, RAFFAELLO AND DONATELLO: THE ITALIAN LOGISTICS MODULES

The start of activities also sees the beginning of the use of the Italian Space Agency's Multipurpose Pressurized Logistics Module (MPLM), built by Alenia Aerospazio and supplied to NASA in three units called Leonardo, Raffaello and Donatello. Carried in the shuttle's cargo bay, the MPLM can transport 9 tonnes of payload to the station, including various instruments and materials, in a pressurized environment. Sixteen active or passive racks can be arranged along its walls; when these are active, power must be supplied by the shuttle.

Nine MPLM missions are planned during the assembly of the station, plus five missions a year thereafter. The typical MPLM mission would last 16 days, 12 with the MPLM attached to the station.

The operational scenario is as follows: When the MPLM arrives near the base, the robotic arm lifts it out

of the shuttle's cargo bay and places it at the attachment point. When attached to the station, it acts as a proper module that can accommodate astronauts, thanks to an environmental control system. At the end of the mission, the arm makes a reverse maneuver, placing the MPLM back in the cargo bay to be taken back to Earth loaded with materials developed on the station.

The hollow metal walls of the MPLM are of 3.2-millimeter-thick aluminum. The cylinder is wrapped in heat-shielding material, over which is antimeteorite shielding. When the MPLM is attached to the station, it receives power from it to activate its data links and communication functions and other operations.

SOLAR POWER

All the power the station requires comes from the Sun, generated by eight flexible solar panels. Each panel is 30.5 meters long and 11 meters wide, with a total surface area of 2,500 square meters. There are 262,400 silicon cells (each of which is 8 centimeters square) on the panels, and each cell supplies 1 watt, so that the total power theoretically available is 246 kilowatts—enough to power 200 homes for a year. However, because of various drains on the system, the actual available power is only 78 kilowatts.

This entire eight-panel "photovoltaic center" was built by Lockheed Martin for a cost of $450 million. The panels are equipped with electromechanical systems able to track the Sun automatically. However, during each 95-minute orbit, the station is in shadow for half an hour when the Earth is between it and the Sun. When this happens, nickel-hydrogen batteries come into action, which recharge when the base is again illuminated by the Sun. A continuous 160/120-volt current runs throughout the station.

THE STATION'S BRAIN

The station is a complex machine that must simultaneously carry out many activities, sustain the lives of the crew and travel safely while maintaining the correct position. To do this, its 40 computers will store the information collected by 2,000 sensors located at strategic points.

Two computers in the laboratory module maintain

the correct flight orientation. Another 14, located outside, are to control other important functions, such as electricity, correct alignment of the solar panels, collecting and expelling heat generated by the internal environment and its devices, and activities of the Canadian robotic arm as it moves up and down the central truss placing or replacing apparatus. Another computer will collect all the data from the sensors and instruments, put it in order and translate it appropriately. A fiber-optic network will allow large quantities of information and images to be transmitted; it will also be more reliable, since it is less subject to electromagnetic interference.

The operating language chosen for the station is ADA, normally used in military systems because of the

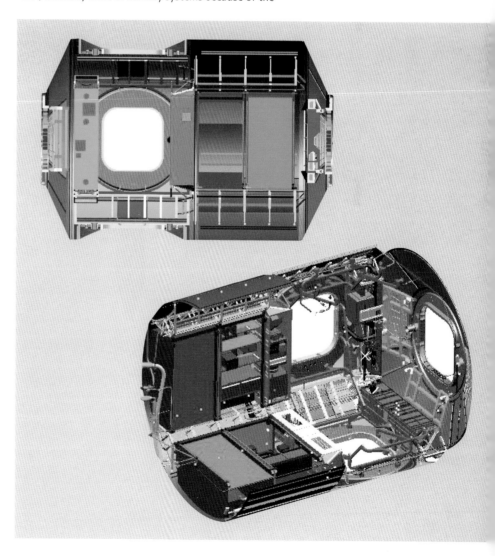

Pictures of Leonardo, the Italian Multipurpose Pressurized Logistics Module (MPLM), during construction and testing at Alenia Aerospazio in Turin.
Far right, bottom: the research ministers from the countries participating in the station project together with the heads of the respective space agencies. Fourth from the right in the front row is Italian research minister Sergio Berlinguer beside ESA director general Antonio Rodotà (third from right).

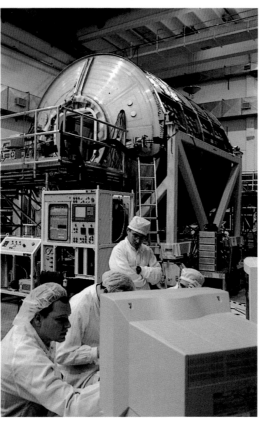

wide range of possibilities it offers. To give some idea of the system's complexity, the flight software alone contains over a million lines of code.

PILOTING THE STATION

Keeping such a large structure in the correct position is not an easy matter. Two systems will do the job: one on the American section and one on the Russian. Both use gyroscopic platforms and navigation satellites—GPS for the Americans and Glonass for the Russians. The latter will also use stellar, solar and horizon sensors together with magnetometers and accelerometers.

Both systems have their own software to handle the data and any corrections that are made by the Americans, using four large gyroscopes with 280-kilogram steel rotors and oriented by electric motors installed in a structure above the laboratory. The Russians have a similar system, with additional rocket engines to control altitude reboosts.

Because of the slight atmospheric resistance the station meets in its journey around the Earth (at an

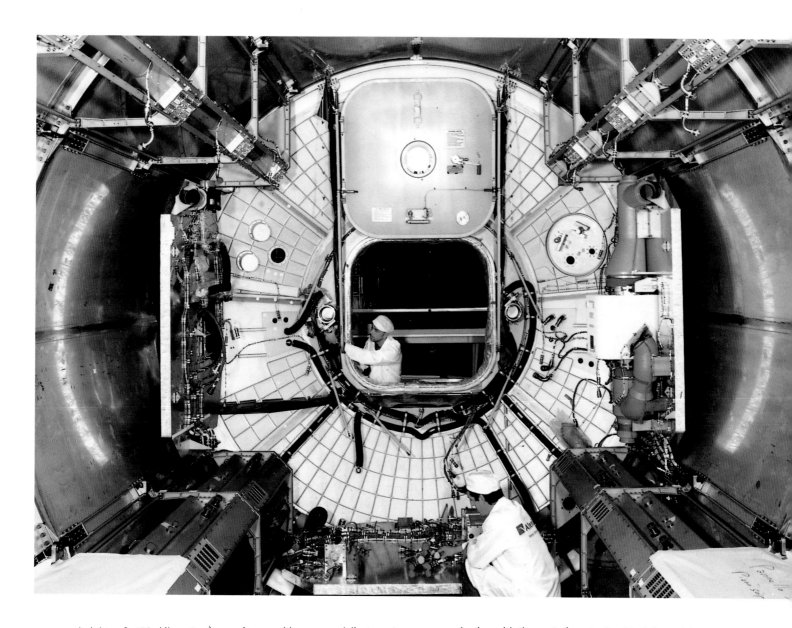

average height of 400 kilometers), gravity would cause the station to slowly descend. So four times a year, the rocket engines will be fired for an orbital reboost. For this operation, however, it is preferable to use the engines on Russian or European vehicles docked with the station. Naturally, the Russian and American computers will "talk" to each other to coordinate their actions before starting anything.

COMMUNICATION AND SURVEILLANCE
Although the station has been designed to have an autonomous orbital life, with built-in duplication of the more delicate systems, communication with the control centers in Houston and Korolev and all the other centers involved in the missions is obviously fundamental. For this, American, Russian, Japanese and European TDRS relay satellites are used to carry voice and video transmissions and the flood of data generated by the instruments. Transmission is at two frequencies: one allows communication with Earth at 192,000 bps (bits per second) and reception at 72,000 bps; the other is reserved for data transmission and functions at 150 million bps.

External control of the station is through four cameras that can be moved as needed to 14 different

Interior of the Italian module Leonardo during assembly.

Illustration of the Canadian
robotic arm that moves up
and down the station's
central beam.

points equipped with attachments and fiber-optic connections. The same network permits internal communication; the astronauts are supplied with portable receiver/transmitters.

THE BEAM: THE STATION'S BACKBONE

The backbone of the station is a trellised truss made mainly of aluminum. It is divided into five pieces, placed in orbit separately and then joined together

with the aid of the Canadian robotic arm and the astronauts. When the elements are launched, they already have all the electronic, heat-control and power-transmission systems they need. At the ends of the truss are the solar panels, and the main body has four attachment points for scientific payloads and research tools. Each point can support a single large 5-tonne, 4-meter-long load or six smaller loads. The attachment points have the necessary electrical power (3 kilowatts at 120 volts) and enable data transmission or storage onboard. Two more attachment points are planned for the unpressurized modules, but they could also be used for additional scientific equipment.

THE RUSSIAN SECTION

What has been described thus far, apart from the Russian FGB and service module, has related mainly to the American elements. Next to be added would be a Russian element—a specialized module like the FGB that allows multiple attachments. Called the Universal Docking Module, it fits into the forward part of the service module and has five docking points at the other end to allow the space settlement to be enlarged.

This is quite important in the architecture of the Russian section, because it indicates where the complex of manned areas will be developed, principally for scientific activities. Two modules equipped with environmental systems and a cosmonaut reentry vehicle are also to be added here. Furthermore, at one of the ports is another cylindrical compartment, similar to that installed on Mir, to facilitate docking with the American shuttle.

The two planned Russian research modules will have similar dimensions and characteristics to those already successfully employed on Mir. What is more, the Russians have proposed transferring the last two Mir modules, Spektr and Priroda, to this area.

A SCIENCE TOWER

At the opposite end of the Universal Docking Module, a trellis tower, pointing upwards, is to be added in front of the service module. At its tip, a set of solar panels—eight at completion (four on each side)—would provide about 25 kilowatts of power. The tower, called the Sci-

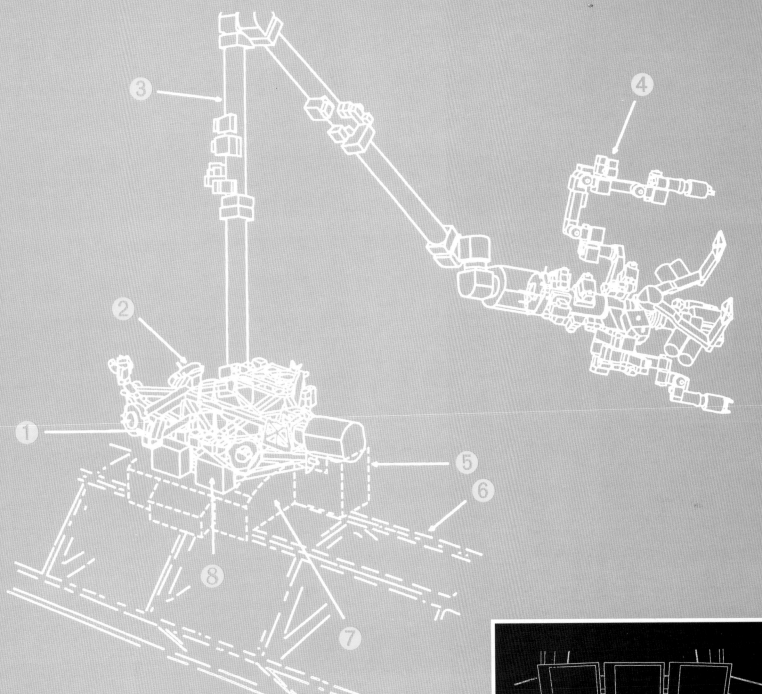

The Canadian robotic arm is 20.5 meters long and moves up and down the station's beam. It is composed of the following elements:

1) TV camera

2) data and power hookup

3) main part of the Space Station Remote Manipulator (SSRM) arm

4) Special Purpose Dextrous Manipulator (SPDM), the "hand," for more precise movements

5) batteries

6) station beam, the arm's transport system

7) electronic apparatus

Right, drawing of the control panels inside the station for the robotic arm.

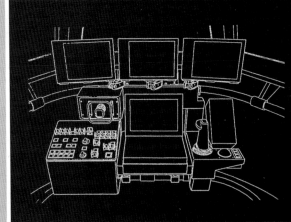

Illustration of the Canadian robotic arm at rest on the central beam of the station and (inset) a detail of the arm at work.

ence Power Platform, is to be built by the Russians and taken into orbit by the American shuttle. It would accommodate scientific instruments that can benefit from the nearby power source. To aid these installations, the tower will be equipped with the European Robotic Arm (ERA) built by the European Space Agency. This allows the various devices located on the tower to be manipulated, removed and replaced. The 8-meter-long ERA is capable of transporting loads of up to 8 tonnes and can provide power and data/video transmission. The arm is designed to be manipulated from outside by an astronaut during a space walk or from a command station inside the Russian service module.

SECOND AND THIRD HABITABLE NODES

An important step in the development of the American section is the arrival of Node 2, to be joined to the front of NASA's laboratory module. Originally, Nodes 2 and 3 were to have been identical, but in 1997, two things happened to change the original plan. Responsibility for this construction was transferred from Boeing to Alenia Aerospazio in Italy. Meanwhile, at NASA, there was a proposal to replace the originally planned habitation module with a new inflatable one so as to experiment with technologies to be used on a future voyage to Mars.

Consequently, some installations from the habitation module were transferred to Node 3 and others to Node 2. At the same time, part of the design was changed, and the two nodes, while retaining the same 4.5-meter diameter, were lengthened to 7 meters (with 3.2-millimeter-thick walls). Each node has two sections of approximately equal length: the first contains the four lateral ports; the second derives from an element of the Italian logistics module.

Node 2 is of particular importance, because it forms

a sort of strategic crossroads for the station, acting as a connector for the Japanese and European laboratories, the centrifuge module and periodic visits by the logistics module. In addition, the American module is to be attached at the rear, and the shuttle would dock at the front using an adapter tunnel. Therefore, Node 2 contains apparatus to supply power to the surrounding modules, as well as all the connections for data transmission and communications and links with the environmental systems.

Node 3, located under Node 1 where the habitation module was to have been, then becomes the habitation module itself. Inside it, besides systems for renewing the air and controlling internal pressure, will be two important installations—the electrolytic plant to generate oxygen from water, and the plant for the treatment and recycling of water—as well as cabins and a shower for the astronauts.

As the large habitation module was eliminated, other astronaut cabins will be placed in Node 2 and the American laboratory. Naturally, Node 3, like the other nodes, will serve as a conduit for fluids, power and data transmission. Finally, it represents an expansion point for the future development of the station.

JEM: THE JAPANESE MODULE

The completion of Node 2 will open the way for the arrival of the Japanese Experiment Module (JEM), to be attached laterally. This is a four-part structure composed of a pressurized laboratory module—similar to the American one and longer than the European one—a logistics module attached at the upper end, an external platform for exposing experiments to the

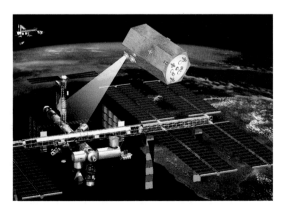

vacuum of space and a robotic arm for working on the platform. While the long laboratory module will offer Japanese astronauts an ideal site for research, the logistics module will allow them to store materials and instruments in eight lockers and transfer these to and from Earth.

The laboratory can accommodate 23 racks (half of them ceded by contract to NASA) and requires 23 kilowatts of power. Normally, it can accommodate two astronauts, but this can be increased to four. It is to be operated with a 32-bit computer and a high-speed data-transmission system that can achieve a maximum of 100 million bps. The planned experiments will relate to life sciences, materials science, research in space engineering and research into technologies to improve conditions for humans in space.

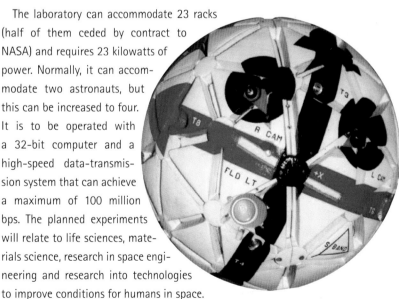

The last element, the robotic arm, has two segments: a 9.9-metre-long section and a shorter, 1.7-meter-long section that can manipulate the 10 experiments carried on the external platform. Its maximum manageable load is 7 tonnes.

Two robotic mini-vehicles capable of flying around the station for inspections. Left and below, NASA's ball-shaped Aercam Sprint with a diameter of 33 centimeters and weighing 16 kilograms. It is equipped with two TV cameras and 12 nitrogen thrusters that move it at 0.03 meters per second, remotely guided by the astronauts. Bottom left, INSPECTOR, the small robot built by the German space agency.

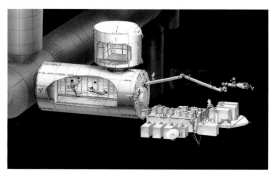

The Japanese Experiment Module (JEM) with an image, above, of the instrument racks installed on the module's two side walls.

THE CENTRIFUGE MODULE

To be able to conduct a series of experiments in a field known as gravitational biology, a module has been built (with the same dimensions as the American laboratory) to house a centrifuge. Initially, it was to have been built in the United States, but in a 1997 barter, the work and responsibilities were taken on by Japan to compensate for the cost of transporting the JEM into orbit.

The aim of gravitational biology is to study how gravity affects life on Earth, from physical development to psychology, and how gravity has influenced human evolution in general. To do this, it is necessary to start with a zero-gravity environment, where a centrifuge can generate gravitational forces of varying intensity, apply them on biological organisms as needed and then analyze the effects.

To that end, the Centrifuge Accommodation Module, located in the upper docking system of Node 2, is to house a 2.5-meter-diameter centrifuge that turns at various speeds to produce gravity ranging from 0.01g to 2g (twice as much as on the Earth's surface). Spinning the centrifuge at 28 revolutions per minute can generate 1g.

Inside will be six different habitats: one for animals, one aquatic, one for insects, a unit for cell cultivation, an egg incubator and a unit for plant research. When it is in operation, five communication channels will be active, each of which can carry 50 million bps (the equivalent of 2,000 pages of text per second). Thus, what is happening in the habitats can be controlled, and the generated data gathered in real time. In the station-assembly sequence, the centrifuge module will be one of the last to be launched.

COLUMBUS (COF): THE EUROPEAN LABORATORY

The ESA will have its own autonomous station element—the Columbus Orbital Facility (COF) module. This is the same size as the Italian logistics module, and part of it was also made in Italy (the structure and heat-control system were built by Alenia Aerospazio, while the German DASA company headed the project). When the COF is connected permanently to Node 2, the node's systems can supply it with the necessary 120-volt electric current, verify atmospheric purity and ensure data transmission to Earth. The COF's autonomous environmental control system will allow astronauts to work in shirtsleeves.

The walls of Columbus are aluminum like those of the logistics module, but they must be thicker (4.8 millimeters rather than 3.2 millimeters) because the COF faces higher environmental risks as it remains attached to the station. The outside is covered with heat shielding overlain by an antimeteorite shield.

The internal fittings are the same as in the other modules: the ceiling and two side walls house experiment racks, and under the floor are operating systems and a storage rack. Racks can be easily switched

The rescue vehicles

The station is equipped with rescue vehicles that act as lifeboats for the crew. They are built by the various countries participating in the project. Together, NASA and ESA developed the Crew Rescue Vehicle (CRV).

Illustrations 1 and 2 show the American version, X-38. The European version, derived from the abandoned Hermes project, is shown in picture 3.

X-38 is 8.5 meters long and 4.5 meters wide and has an attitude-control system of rocket engines fueled by nitrogen.

According to NASA's plan, the CRV has to be capable of transporting seven passengers and returning to Earth automatically or piloted by astronauts. It is powered by onboard batteries and guided by the GPS satellite system. It has autonomous operating capacity of 9 hours and docks at the lower part of the station (4).

Russia, on the other hand, continues to use the tested and proven three-person Soyuz capsules (5).

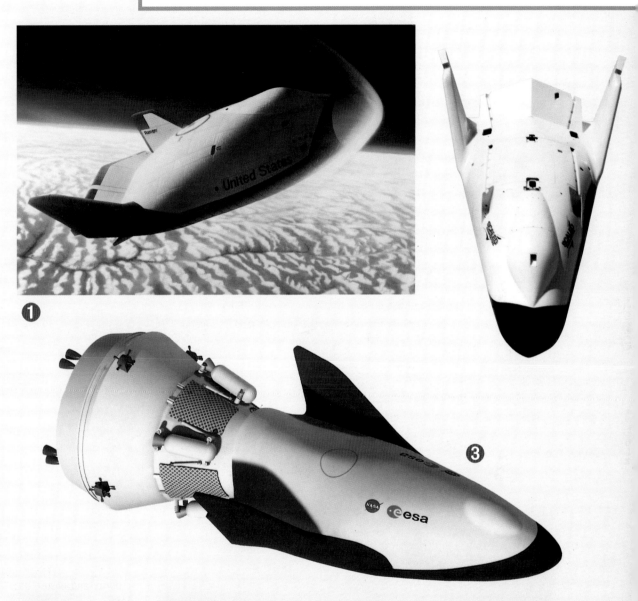

❶

❸

❺

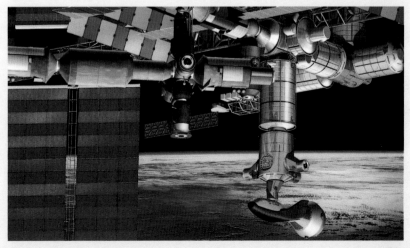

❹

depending on the experiments to be run. They can also be exchanged with the other laboratories, as they have been built according to the same specifications.

The COF can accommodate 10 research racks, five of which will be used by NASA according to agreements governing the right to use station resources. The five reserved for ESA contain equipment for experiments in the Microgravity Facilities for Columbus (MFC) program, prepared by ESA or by national agencies. They are dedicated to research in physiology, fluid science, biology (Biolab) and materials science. The fifth rack is for storage.

These racks are "active," having electrical hookups, connections for receiving and sending data and heating and cooling systems. In addition, three "passive" racks are reserved for storing materials and astronaut equipment.

On its forward exterior, the COF has attachment points like small balconies, where experiments requir-

ing vacuum conditions can be placed. Management of all the activities on Columbus will be done by a "brain" developed by Matra and called the Data Management System, which will run all the different jobs and govern data gathering and transmission traffic. Each component of the European module is designed to operate safely for at least 10 years.

ANTIMETEORITE AND DEBRIS SHIELDS

All external surfaces of the station modules require protection from a threat that has grown in the past few decades as activities in space (such as orbiting satellites) have increased; that is, the hazard posed by debris fragments of various sizes drifting in space around Earth. Added to these are natural meteoroids, also of varying dimensions. Statistically, however, it is the man-made objects that are the most dangerous.

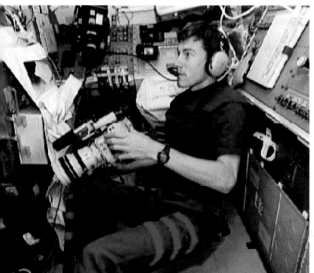

The first crews chosen for the International Space Station: Yuri Gidzenko, left; Sergei Krikalyev, far left; Bill Shepherd, first commander of the base, left center; and Susan J. Helms, selected as a member of the second crew, below.

Bottom, from left to right: Kenneth D. Bowersox and Vladimir M. Dezhurov (chosen as members of the third crew); Daniel W. Bursch and Carl E. Walz (chosen as members of the fourth crew).

The two large drawings on this and facing page show positions spontaneously assumed by the human body in zero gravity on the space station.
Below, three views inside the American part of the station, from top to bottom: astronaut's personal cabin, shower and kitchen area with a microwave oven.
Right, footrest used by the astronauts to help them hold position while working.

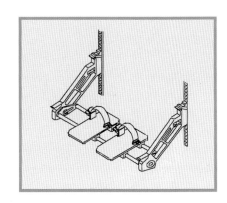

NASA asked the U.S. National Research Council to look closely into the nature of the threat which the station would face in its orbital life. In 1997, a committee set up within the council for this purpose published a report that not only examined and evaluated the risks but also suggested ways to give adequate protection to the orbital base and, above all, to the people living on it. First, the committee estimated that the risk of impact with meteoroids or space debris was one of the 15 most serious risks for the station. The probability that the base would be hit by a "critical object" capable of damaging it was determined to be 10 percent over 10 years.

It was consequently decided to build protective shields able to defend the base from objects 1 centimeter in diameter, to plan evasive maneuvers to avoid objects larger than 10 centimeters wide (which could also be detected by terrestrial radar) and to adopt measures to alleviate damage caused by objects with a diameter between 1 and 10 centimeters.

Studies aimed at evaluating the environment where the station would have to operate (conducted in the 1990s by the U.S. Space Command) demonstrated that 99 percent of the threat would come from particles small enough to be neutralized by adequate shielding. In estimating the risks, however, it was expected that from 1 to 10 evasive maneuvers would be needed each year to escape impacts with fragments having a diameter more than 10 centimeters.

Various theoretical models developed (the last in 1996) used data gathered from shuttle flights, the LDEF satellite (that remained in orbit for 68 months before being recovered) and the Russian experience with the Salyuts and Mir. It was decided to build an external shield for the station that could protect it from objects equivalent to an aluminum sphere 1 centimeter in diameter traveling at an average speed of 12 kilometers per second. The result was protection that varies in different spots according to the perceived degree of threat (for example, the two ends were deemed to be least exposed to risk). In general, shielding consists of an outer layer of aluminum about 1 millimeter thick. Under this are two layers of

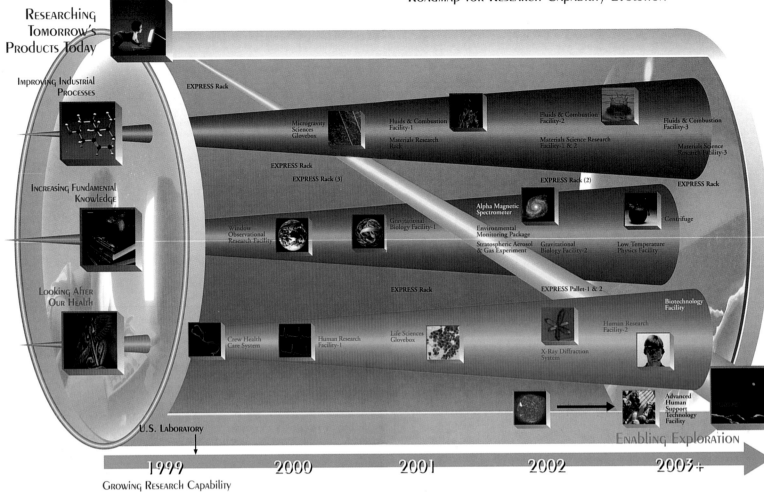

RESEARCHING
TOMORROW'S
PRODUCTS TODAY

IMPROVING INDUSTRIAL
PROCESSES

INCREASING FUNDAMENTAL
KNOWLEDGE

LOOKING AFTER
OUR HEALTH

EXPRESS Rack

Microgravity Sciences Glovebox
Fluids & Combustion Facility-1
Materials Research Rack
Fluids & Combustion Facility-2
Materials Science Research Facility-1 & 2
Fluids & Combustion Facility-3
Materials Science Research Facility-3

EXPRESS Rack
EXPRESS Rack (3)
EXPRESS Rack (2)
EXPRESS Rack

Window Observational Research Facility
Gravitational Biology Facility-1
Alpha Magnetic Spectrometer
Environmental Monitoring Package
Stratospheric Aerosol & Gas Experiment
Gravitational Biology Facility-2
Low Temperature Physics Facility
Centrifuge

EXPRESS Rack
EXPRESS Pallet-1 & 2
Biotechnology Facility

Crew Health Care System
Human Research Facility-1
Life Sciences Glovebox
X-Ray Diffraction System
Human Research Facility-2
Advanced Human Support Technology Facility

U.S. LABORATORY
ENABLING EXPLORATION

1999   2000   2001   2002   2003+

GROWING RESEARCH CAPABILITY

Right, ground simulations for the astronauts at Houston using virtual-reality tools. Above, NASA chart showing the fields of research in which the astronauts will work on the station.

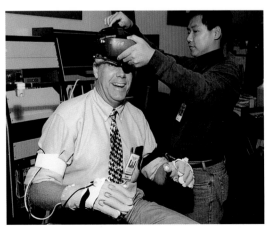

fibrous material (Nextel and Kevlar) able to absorb the fragments created by the initial impact with the outer aluminum layer before they reach the aluminum walls of the modules, whose thickness varies between 3.2 and 4.8 millimeters (greater in areas deemed to be most at risk).

THE STATION ENVIRONMENT

The habitable volume inside the station is equal to that of two Boeing 747 jumbo jets. The American section has its own life-support systems, as does the Russian.

However, the internal atmosphere (nitrogen and oxygen in terrestrial proportions) is equalized between the two sections, and the door between them is always open. Therefore, an environmental failure in one of the areas could be compensated by the other. Temperature can be regulated to around 20°C, and pressure is the same as at sea level. Heat generated inside is expelled into space by circulating water through the external radiators.

USE OF THE STATION

The birth of the International Space Station represents a great opportunity for the countries involved in its construction as well as for those who hope to conduct research onboard. In outer space at an altitude of 400 kilometers, the pull of gravity is an average of 10,000 times lower than on Earth's surface. It is therefore virtually negligible, offering the possibility to create substances impossible to obtain in normal gravity on the ground.

The basic element to be used for research activities is the International Standard Payload Rack (ISPR), which has a volume of 1.6 cubic meters and can carry a load of up to 700 kilograms. An Express Rack is equipped to handle smaller experiments in a shorter time. All the racks conform to the American standard, except for the ones in the Russian section, where nonstandard containers are used.

According to the participation agreements made at the outset of the project, the United States will use 97 percent of its own laboratories and 46 percent of each of the European and Japanese ones. NASDA and ESA will enjoy 51 percent of the use of their own laboratory modules. Canada has the right to 3 percent of all the research space, while Russia will use 100 percent of its modules. The scientific activity to be done has been divided into 11 areas: biotechnology, combustion physics, fluid physics, materials such as silicon and ceramics, electronics, alloys and metals, chemistry and polymers, life science and biomedical science,

Some experiments conducted in zero gravity, from left to right: combustion studies and crystal growth.

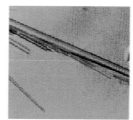
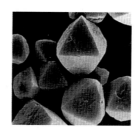

The suits for the astronauts' space walks

Two types of space suit are used on the station: the American Extravehicular Mobility Unit (EMU) and the Russian Orlan-M. Both are improved versions of earlier models. The EMU was used on the shuttle, and the Orlan on Mir. The refinements concern safety, prolonged use and details that improve performance, such as widening the visor.

The American suit, made up of two parts joined at the waist, can be used for 25 EVAs and remain in orbit for 180 days before being checked. It can be adapted to the astronaut wearing it and allows EVAs of 7 hours plus a half-hour emergency leeway.

The Russian suit, a single piece put on from the back, can be used for 10 EVAs and a 4-year term in orbit. It is capable of 5 hours of autonomous operation plus a half-hour emergency leeway.

Both supply the astronauts with oxygen and have a similar weight (117 kilograms for the U.S. model and 110 kilograms for the Russian). The American suit also has a "life jacket" mini-rocket pack, called Safer, which is added to the bottom of the normal pack to allow greater movement in emergency conditions. Safer has also been adapted for the Russian suit.

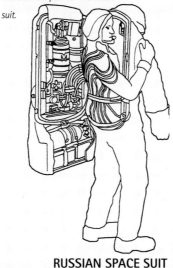

RUSSIAN SPACE SUIT

AMERICAN SPACE SUIT

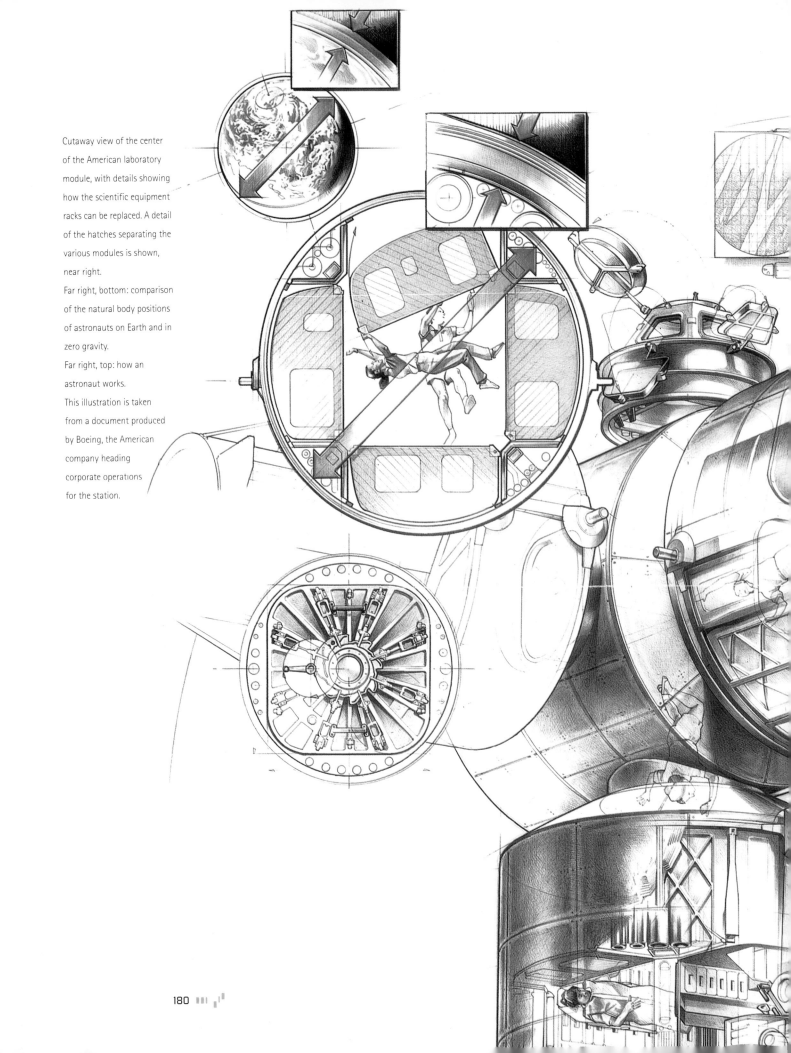

Cutaway view of the center
of the American laboratory
module, with details showing
how the scientific equipment
racks can be replaced. A detail
of the hatches separating the
various modules is shown,
near right.

Far right, bottom: comparison
of the natural body positions
of astronauts on Earth and in
zero gravity.

Far right, top: how an
astronaut works.

This illustration is taken
from a document produced
by Boeing, the American
company heading
corporate operations
for the station.

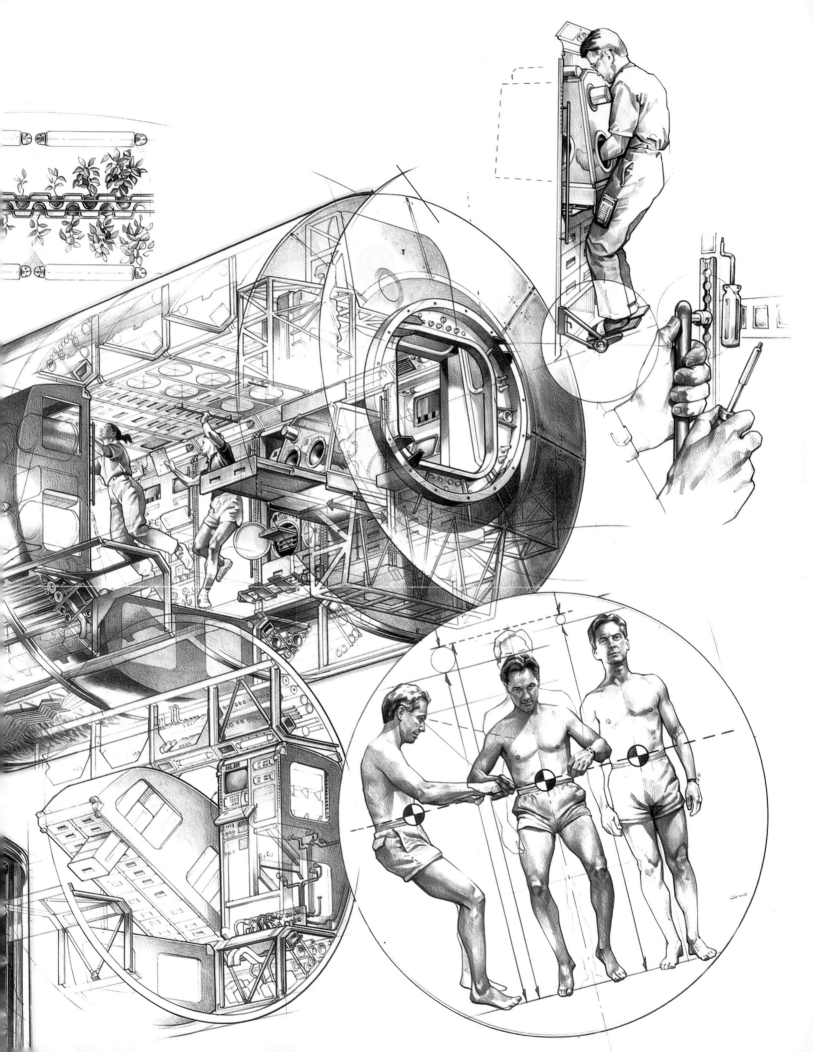

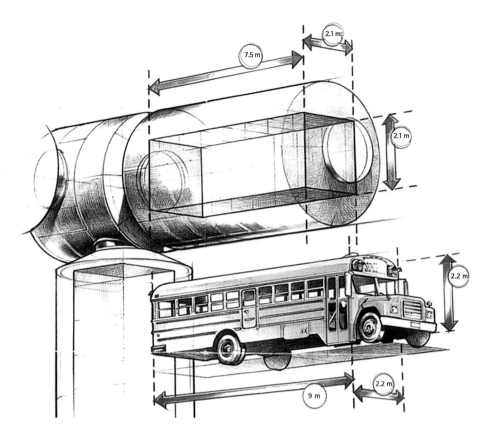

7.5 m

2.1 m

2.1 m

2.1 m

2.2 m

2.2 m

9 m

Comparison of the size of the station's American laboratory module and a typical school bus.

engineering research and the development of technology, investigating commercial applications (from robotics to power generation and remote sensing of the Earth's surface), and observation of the atmosphere and the terrestrial environment.

The work is to be organized into "research campaigns" lasting from 30 to 365 days and taking into account not only the available resources but also the selection of astronauts and their training. To support the campaigns, five shuttle flights will be made each year, plus Russian, European and Japanese launches.

According to NASA, every dollar invested in space gives a return of $9 in new products, processes and technologies on Earth. Thus, space activities are to be viewed as an investment for the future as far as economic aspects are concerned. But to this should be added the development of knowledge, which supports the growth of a country from all points of view, contributing to its wealth on a broader level.

For this reason, in the second half of the 1990s, the space agencies involved in the station (America's NASA, Russia's RKA, Europe's ESA, Italy's ASI, Japan's NASDA, Canada's CSA and Brazil's BSA) launched programs to stimulate the interest of the scientific and

corporate communities, and thereby the arenas of public and private research, in preparing research projects to be conducted onboard. Meanwhile, plans are also under way for commercial use of the station once it has been built.

STATION INHABITANTS: THE FIRST FOUR CREWS

In November 1997, NASA and the RKA selected the first four crews that would set foot on the new International Space Station after assembly of the Russian FGB and service module and the American Node 1. The first three inhabitants chosen were Yuri Gidzenko and Sergei Krikalyev from Russia and Bill Shepherd of the United States, who was named the station's first commander. Gidzenko was selected to command the Soyuz capsule in which the trio would be launched from Baikonur. The choice of an American as the base's first commander created some dissent among the Russian cosmonauts.

It was decided that the first crew would stay onboard for five months, during which time they would help assemble the next elements (the U.S. laboratory module, the tower for the first solar panels and Canada's robotic arm) as well as monitor the functioning of the systems.

William M. (Bill) Shepherd, a U.S. Navy captain, was born on July 26, 1949, in Oak Ridge, Tennessee, and became an astronaut in 1984. His flight on the orbital base would be his fourth space experience, as he had previously flown 440 hours on the space shuttle on three missions in December 1988, October 1990 and October 1992. Shepherd knows the station extremely well, because he has also been vice administrator for the project.

Yuri Pavlovich Gidzenko, an air force colonel, was born in the village of Elanets, Elanetsky district, in the region of Nikolayev. He became a member of the cosmonaut corps in 1989 and spent 180 days on Mir during ESA's Euromir mission from September 1995 to February 1996.

The third member of the crew, flight engineer Sergei Konstantinovich Krikalyev, born in St. Petersburg on August 27, 1958, is perhaps the most famous

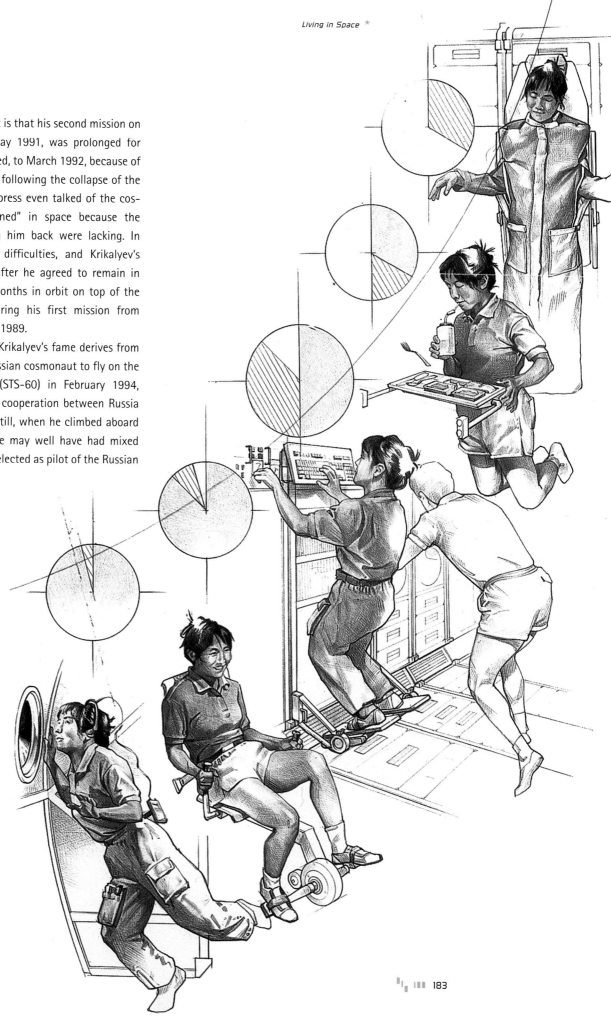

for two reasons. The first is that his second mission on Mir, which began in May 1991, was prolonged for much longer than planned, to March 1992, because of problems on the ground following the collapse of the USSR. At the time, the press even talked of the cosmonaut being "abandoned" in space because the means to go and bring him back were lacking. In truth, there were real difficulties, and Krikalyev's sojourn was extended after he agreed to remain in orbit. So he spent 10 months in orbit on top of the five he had logged during his first mission from November 1988 to April 1989.

The second source of Krikalyev's fame derives from having been the first Russian cosmonaut to fly on the U.S. shuttle Discovery (STS-60) in February 1994, opening the new era of cooperation between Russia and the United States. Still, when he climbed aboard the American shuttle, he may well have had mixed emotions: he had been selected as pilot of the Russian

A day in the life of an
astronaut on the station
(from top to bottom):
1) sleep, 8 hours
2) food and hygiene, 4 hours
3) work, 9 hours
4) physical activity, 2 hours
5) free time, 1 hour

The automatic Progress, HTV and ATV supply ferries

In addition to NASA's shuttle that will carry the Italian modules Leonardo, Raffaello and Donatello in its cargo bay and then dock with the base, the station is to be supplied by three automatic vehicles: Russia's Progress M-1, Japan's Hope Transfer Vehicle (HTV) and Europe's Automated Transfer Vehicle (ATV).

The Russian capsule is 7.5 meters long and carries 2.7 tonnes of supplies (water, oxygen, nitrogen, fuel and various materials). Its engines can also be used to boost the station's orbit. It has autonomous flight capability of 30 days, and once the mission is over, it is destroyed in the atmosphere.

NASDA's HTV, launched with the Japanese H-2 rocket, is shaped like a shuttle, and its payload is carried in a pressurized container in the cargo bay. Once it gets near the base, it does not dock but is captured by the robotic arm. Unlike all the other vehicles, the HTV carries no fuel.

ESA's ATV is 8.5 meters long and is launched by an Ariane 5. It can transport 4 tonnes of fuel and 5 tonnes of various supplies. The ATV can remain attached to the station for up to 6 months and can also be used to boost the base's orbit.

HTV

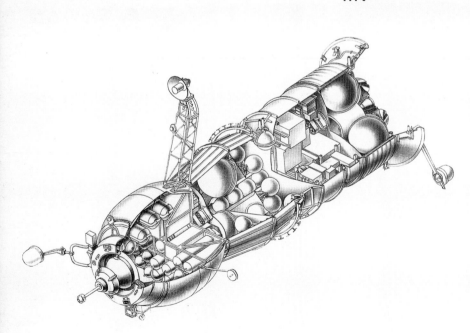

PROGRESS

ATV

Buran shuttle, but he never had the chance to sit at the controls because it never made a manned flight.

His mission on the new station would therefore be his fourth experience in space. He already has a total of 15 months of orbital life behind him, during which he made seven space walks. All in all, Krikalyev is one of the best prepared cosmonauts in the Russian team.

One of the accepted policies, perhaps as a result of the Russian protests over the choice of an American as first station commander, is that of alternation, meaning that a Russian would follow an American as commander. So when the shuttle Atlantis docks with the orbital base after five months, the plan is that it will be carrying the second crew, led by Yuri Usachev as commander, with Americans James S. Voss and Susan J. Helms.

Helms was born in Charlotte, North Carolina, in 1958. She is a colonel in the USAF and an aeronautical engineer with a Master's degree in aeronautical and astronautical sciences from Stanford University. All the crew have space-flight experience, including Susan Helms, who previously took part in two shuttle missions for a total of 406 hours.

The European Columbus Orbital Facility (COF) laboratory module built by ESA, and the wall lockers with the scientific instruments.

From top to bottom: the station seen from below with a rescue vehicle docked at the bottom; illustration of ESA's science platform on the station's beam; sectional view, below, of the COF module; Europe's COF module, bottom right, attached to Node 2, together with the centrifuge module.

Facing page, NASA's shuttle docked with the station (the Italian MPLM Leonardo is in the cargo bay).

The third crew is to be commanded by Kenneth D. Bowersox and includes two Russians, Vladimir Dezhurov and Mikhail Turin. The first two have flown in space before, while Turin would be making his debut. The fourth trio selected consists of Yuri Onufrienko (commander), Carl E. Walz and Daniel Bursch.

During the assembly phase, the large base is to be occupied by crews of three. Once it is completed, however, when all the modules and resources are available, that number will rise to six and sometimes even seven, with at least three or four crew members working solely on research. Stays will vary from three to six months.

The inhabitants' daily schedule and organization of work will be different from those on the American shuttle missions. There will no longer be a rigid roster of schedules and work predetermined on Earth. Instead, the crew will run itself, as was already partly the case on Mir. This will be possible because the base will by then have achieved a sufficient level of autonomy and security. In addition, because of the introduction of "telescience," allowing experiments to be

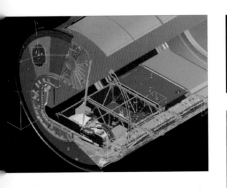

The station seen from below, showing the TransHab module that will test technologies for a trip to Mars attached to the central section.

controlled directly from ground centers, the situations to be faced will change continuously. Life on the international station will therefore take on a more natural rhythm, similar to that to be found, for example, at a scientific base in the Antarctic.

FROM THE STATION TO MARS?

At the end of 1997, there was a surprising new development in the station project. The American magazine *Aviation Week & Space Technology* revealed the existence of a NASA project to construct a new type of habitation module replacing the one already under construction. This habitation module would be the last stage in the assembly of the orbital base.

The project, known as TransHab, was decidedly innovative, involving a module that would actually be inflated in orbit. Its inside volume would be three times larger than the original habitation module, and it would be attached to Node 3. TransHab was designed with Mars in mind. Its design would permit both a trip to the Red Planet and a stay on its surface. The plan represents a formidable engineering challenge, and using the module on the station would serve to test technologies that could in future be able to achieve the objective of a Mars landing. Among these is the new realm of electronic nanotechnology.

TransHab is a cylinder 7.5 meters in diameter and 8 meters long. Carried into orbit in the shuttle's cargo bay, it is to be inflated in orbit before being attached to the station. Its overall weight, 4.5 tonnes, is half that of the metal module previously designed, thanks to the fact that TransHab's walls are made of Vectran, a material similar to Kevlar but with better structural properties. Externally, TransHab is protected from micrometeorites and space debris by four layers of Nextel (like the other modules) plus heat shielding.

Its life-support systems are equally innovative. They use a biological system to recycle water and a combined biological-mechanical system to renew the air. An apparatus containing cultivated wheat and fed with carbon dioxide will supply 25 percent of the crew's oxygen needs. Urine is recycled, while solid waste is incinerated to provide carbon dioxide for the cultivation.

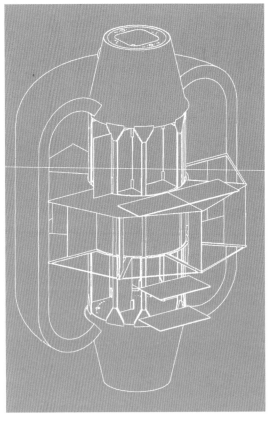

Drawings of the TransHab habitation module designed by NASA to experiment with technologies necessary for a voyage to Mars.

As we begin to look toward Mars, assembly of the International Space Station has begun. The first two modules, Zarya and Unity, were connected in orbit in December 1998, forming the seed for the first real space base able to support a relatively normal life without the conditions of daily experimentation experienced on the Salyuts, Skylab and Mir. It has taken a century for the first visions of Konstantin Tsiolkovsky to be transformed into a permanent base in outer space. But in the end, human evolution has been able to achieve this step too. Now the prospect of using the station as a departure point for an easier and more "natural" exploration of the solar system, as anticipated by Wernher von Braun, also comes closer to reality—a base orbiting the Earth could act as the launch point for an interplanetary voyage, thus avoiding a much more difficult terrestrial lift-off and reentry.

But we are only at the beginning. To be transformed into an infrastructure capable of realizing such a dream, the International Space Station will need to grow and develop capabilities that are as yet only embryonic. Still, the first and most difficult jump has been made, and in a certain sense, Tsiolkovsky's challenge of 100 years ago—when he said, "The Earth is the cradle of mankind, but man cannot live in the cradle forever"—has been taken up. The "cradle" has now moved into orbit around the Earth. Perhaps moving it again, to other planets and, in future, to other stars, will prove to be a more natural step than this first leap beyond our home planet.

Where and how astronauts and cosmonauts are trained

Astronauts prepare for their missions in different types of station simulators on the ground. The U.S. training center is the Johnson Space Center in Houston, Texas, shown in two top photographs and two black-and-white photos at far right, where life-size models of parts of the station have been built. At Johnson's Sonny Carter Training Facility, there is a Neutral Buoyancy Laboratory: a large swimming pool containing station modules. Work to be done during space walks is simulated here with zero-gravity conditions partly reproduced in the water. American, European and Japanese astronauts are trained at the Johnson center. Japan also has its own NASDA training center at Tsukuba, with simulators and a tank, far right, bottom. In Russia, the cosmonauts are prepared at Star City, near Moscow, where the techniques are the same, with mock-ups and a tank, right. ESA does some if its training work in the COF module simulator at Estec in Noordwijk, the Netherlands.

Four representations of the
entire International Space
Station in Earth orbit.

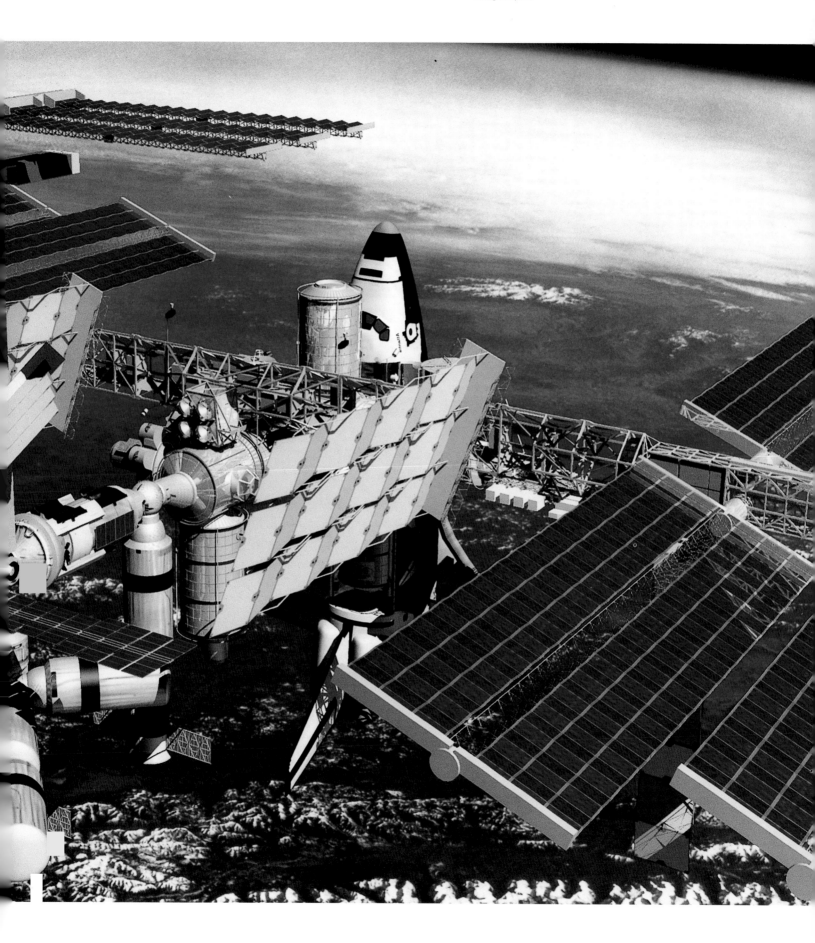

Colonies in space

7

From O'Neill's
designs to the
future

Colonies in space

If it took a century to arrive at the International Space Station, the first established human settlement in space, what does the future hold for space stations? In some ways, the question takes us back to the beginning of this book and Hale's fantastic "brick moon" of 1869. In the intervening decades, scientific and technological knowledge and skills have developed enough to construct a manned base orbiting the Earth. Now science fiction may again help us to imagine what kind of future humans can forge in the cosmos.

Preceding pages: illustration of one of O'Neill's space colonies.

Above: Princeton physicist Gerard O'Neill.

Facing page: illustration of Island One, top; and a detail of the rings in the agricultural section, bottom.

WHEN THE SUN DIES

The human race, born on planet Earth in a solar system near the edge of the Milky Way Galaxy, can continue to survive only if it ultimately abandons its original planet. When our Sun has exhausted its energy resources, it will fade and become a "red giant" star. The tremendous expansion typical of this dying phase of the star will eventually absorb the Earth and inevitably destroy it. Or, according to another theory, even before that, the increased pressure of the solar wind as its density increases will push the Earth away from its original orbit, at the same time wiping out its life forms. Whatever the final scenario, the result does not change. The fate of the human race is sealed if it does not emigrate toward other stars in other solar systems.

All this will happen a few billion years from now: according to astrophysicists, the Sun is about halfway through its life span and should therefore burn for at least another 5 billion years. The figures seem incredible in comparison with the span of a human life. But looked at with a more "cosmic" eye, in a dimension where time, space and even a species take on different meanings, much less limited than the way we measure them on Earth, imagining a future far from our original planet becomes natural because it obeys a fundamental rule of the evolution of any species—the quest for survival.

The space station therefore responds to an inevitable, natural evolutionary drive—it is the first tool humans have constructed to learn to live in the cosmos. It also lays the basis for developing and increasing the capabilities necessary to face the future

voyage outward from Earth. It is the beginning of humankind's new space dimension.

Now the space station—with secure life-support systems, prospects for future expansion and the fact that it is no longer a temporary adventure but a stable place to carry out activities useful to the human race—represents a "home" where the inhabitants can live some sort of life. Thus, a double life will develop for humankind: one on Earth and one on the station. Seen from space, both are stations in the cosmic darkness—Earth can be viewed as a large, natural station capable at a certain point in its history of spawning a smaller, artificial one in orbit around it.

It is therefore logical to imagine a future in which orbital bases will multiply and become bigger and more specialized in their functions. And little by little, they will be placed in different orbits, farther and farther away from the Earth that spawned them.

O'NEILL'S COLONIES

It was from the above perspective that Gerard K. O'Neill of Princeton University used all the tools of science and technology at his disposal in the mid-1970s to examine the fascinating concept of "space colonies."

O'Neill was born in Brooklyn in 1927. In 1954, when only 27 years of age, he was teaching physics at Princeton University. As competent in basic science as in its applications, he first worked on the physics of elementary particles in atomic nuclei. Then he specialized in the technology of accelerators to make those particles collide at high speeds and study the results, which were sometimes other particles, other basic building blocks of matter.

In 1969, while American astronauts were landing on the Moon, O'Neill began examining the prospect of man-made space colonies in various orbits. The energy crisis of the early 1970s highlighted the problem of our planet's finite resources and the need to find alternatives so as not to damage the Earth seriously or destroy it completely. The Princeton scientist identified five issues that in his opinion were crucial to the survival of the Earth and humanity. He said these could be resolved by the creation of colonies in space. The five issues concerned the need to raise every human being to a standard of living then reserved for only a select few, protect the biosphere from the damage caused through pollution by industry and transport, create quality living space for a population that doubles every 35 years, find clean energy resources and prevent the overheating of the environment. The key to all of these, according to O'Neill, was the availability of a clean energy source, and for this, he suggested Satellite Power Stations to be built with material mined from the Moon or asteroids.

In an article published in the journal *Physics Today* in 1974, the Princeton scientist laid out a series of specifications resulting from five years of study. In the succeeding years, a number of conferences and further research supported by Princeton and Stanford universities, the American Association for Engineering Education and NASA's Ames center examined the possible architecture of space colonies in more detail.

In May 1975, at another Princeton University meeting, the term "Space Manufacturing Facility" came into use to describe installations dedicated to space

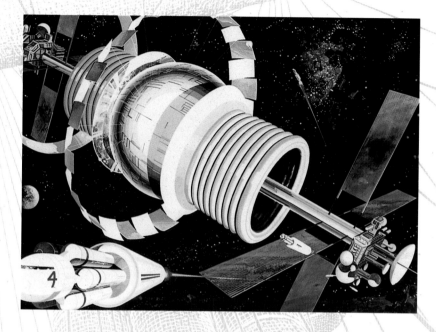

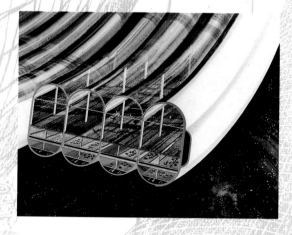

construction activities. Their first objective would be the building of Satellite Power Stations and orbiting bases to house the industrial activities that today pollute the Earth's environment.

THE THREE ISLANDS

In 1976 and 1977, O'Neill published two editions of his book *The High Frontier* in which he gathered together the plans and proposals made in the preceding years. This was an enormous undertaking which probed the diverse and complex aspects of constructing a space colony from an engineering point of view. It was therefore practical, not merely theoretical. For example, O'Neill elaborated a family of colonies with the capacity for expansion. These he called Islands One, Two and Three.

The first was a colony of 10,000 people. The "Island" would be an aluminum sphere with a circumference of 1.6 kilometers (500 meters in diameter). The living quarters would be in the middle zones, up to 45 degrees of latitude north and south of the sphere's equator. Above and below the sphere would be two cylindrical structures devoted to agriculture and composed of a toroidal series of transparent wheels. With sunlight coming through wide windows, the overall effect was that of a gigantic greenhouse.

Interior of the colony, above; and its control center, right. Below, drawing showing operation of the mass launcher designed by O'Neill to transport material in space. The surrounding coils generate a magnetic field that launches the central container.

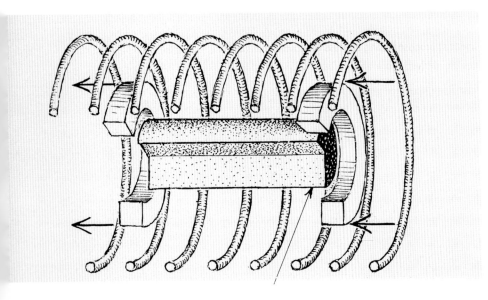

Island One would rotate on its own axis 28 times a minute, creating artificial gravity. According to O'Neill's calculations, 10,000 inhabitants was the minimum number necessary to initiate reproduction and growth in the colony. Inside the sphere, each inhabitant would have 45 square meters of personal space. A family of five would have an apartment of around 230 square meters, with a private garden terrace taking up a quarter of the area.

The living area would be concentrated in the midzone of the sphere, with the rest given over to parks, gardens, commercial areas, streams and free areas for all the inhabitants. A shallow river could run at the equator (where the aluminum walls of the colony were up to 18 centimeters thick), opening out into pools for swimming. The banks would be formed of lunar sand, and farther back, there would be treed areas with cycle, walking and athletic tracks.

The external shields, separated from the rotating structure, would be spherical shells made of lunar material that would protect the inhabitants from dangerous cosmic radiation. Finally, sunlight was to be introduced into the environment using mirrors.

According to O'Neill, after a dozen Island Ones had been built, an Island Two could be begun, even bigger than Island One. Its central sphere would have a circumference of 5,600 meters (1,800 meters in diameter), and the population would be the size of a small city of 140,000.

A greater population than that would require a radical change in architecture. Thus was born the design for Island Three, with 10 million people distributed in towns of 25,000. At this point, there would be colonies with different functions, as O'Neill put agricultural and industrial production on two separate bases close to the residential one.

The architecture of Island Three would be a double cylinder, each one autonomous and rotating on its longitudinal axis twice a minute to create artificial gravity. The cylinders were to be 80 kilometers apart, each being 30 kilometers long with a diameter of 6 kilometers, for a total area of 1,300 square kilometers. The internal atmospheric pressure would be the same as at 1,500 meters on a terrestrial mountain.

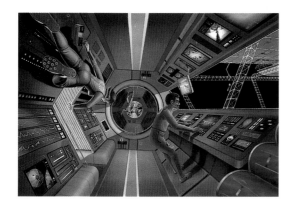

The cylinder walls were to be constructed with transparent windows to let in the sunlight. Nontransparent zones, 3 kilometers wide, would accommodate long ramparts. The light for the interior would be reflected off large mirrors that unfurled outside like flower petals. As they opened and closed, they would mark dawn and dusk, with the length of a day being the same as on Earth. As the longitudinal axis of the cylinders would always point toward the Sun, there would be continuous illumination and the capacity to regulate it as desired.

"With industry and agriculture on the outside," wrote O'Neill, "the inhabitants of Island Three can use their territory as a rest and recreation area. I believe that with the arrival of colonists from various parts of the world, there will be a great variety of ways in which the inhabited part will be used. Some immigrants might choose to populate their part with small villages of single-family houses separated by forests. Others might choose to build densely inhabited, intimate cities to enjoy, for example, the color, excitement and human interaction typical of Italian towns."

There would be no worries about food reserves, because the automated "agricultural colonies" could produce four harvests a year. Naturally, full recycling was to be practiced on all the colonies.

THE WHEEL COLONY

In 1975, O'Neill and a group of 29 experts developed a plan for NASA for a wheel-shaped (toroidal) colony 1,790 meters in diameter, reproducing the classic design proposed by Wernher von Braun in the 1950s.

The 130-metre-wide tube forming the rim of the wheel would house 10,000 people, with artificial gravity generated by rotation (one revolution per minute). Corridors would connect the wheel to the central area where arriving spaceships would dock. In this design, sunlight was to be reflected by a circular mirror placed above the wheel.

O'Neill considered it essential that the colonies be built of materials extracted from the Moon and, later, asteroids. The location he considered optimal for such a colony was at the L5 point of the lagrangian system (named for Lagrange, an 18th-century Italian mathematician)— that is, at one of the five positions in the Earth-Moon system where gravitational forces cancel each other out, creating stable orbital conditions. The L5 point forms the apex of an equilateral tri-

Top left, interior of the colony. Background, drawing of the Island Three colony. Bottom, a solar power station under construction.

angle 384,000 kilometers equidistant from the Earth and the Moon.

To deliver the construction material, the Princeton scientist proposed using a "mass launcher" of his own design, derived from the particle accelerator tech-

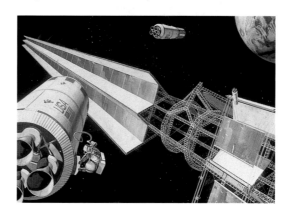

A wheel-shaped colony designed by O'Neill's group to be capable of housing 10,000 inhabitants. Facing page, illustration of the interior of the wheel, with homes and vegetation.

niques in which he was an expert. The materials (aluminum, for example) would be launched at lagrangian point L2 above the Moon's orbit and then transferred to the L5 point, where they would be used to produce the necessary pieces for construction of the colony.

O'Neill's research was done at the same time as that of another American scientist, Peter Glaser, who concentrated his attention on the construction of solar satellites, large automatic platforms able to collect solar energy and transmit it to Earth as microwaves. Glaser also thought that such energy could subsequently be used to power a space colony, which is why O'Neill followed his progress so attentively.

STATIONS AROUND MARS?

After the 1970s, no effort similar to O'Neill's investigated the future of cosmic colonies. One of the reasons for this was that in the interim, the time had come to build the first real, large, secure space station—a project that took 20 years. It was to this end that energy and resources had to be concentrated, and rightly so.

Now that the station is a reality, it is again time to think about the future and the prospect of more complex space settlements around other planets, as has, for example, been suggested for Mars. The starting point could be O'Neill's work, with help, perhaps, from the visions of science fiction.

Appendix

Chapter Contents

The Russian Salyut stations:
Seven military and civilian laboratories

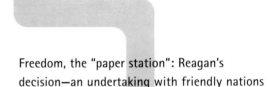

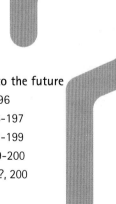

Index of Names

Bibliography

Advisory Committee on Redesign of Space Station. *Final Report to the President.* June 1993.

Amadieu & Heloret. *The European Automated Transfer Vehicle.* IAF-97.

American Astronautical Society. *History of Rocketry and Astronautics.* International Academy of Astronautics Symposia, vol. 8. San Diego, 1993.

American Institute of Aeronautics and Astronautics. *Space Stations and Space Platforms: Concepts, Design, Infrastructure and Uses.* Progress in Astronautics and Aeronautics, vol. 99. New York, 1985.

Ananoff, A. *L'Astronautique.* Librairie Artheme Fayard.

Barth, H. *Hermann Oberth.* Uni-Verlag.

Bekey, I. *Permanent Presence: Making It Work.* AAS, vol. 60.

Belew & Stulinger. *Skylab: A Guidebook.* EP-107-USPO.

Blaine, B.C.D. *The End of an Era in Space Exploration.* AAS, vol. 42.

Bluth & McNeal. *Update on Space.* National Behavior Systems.

Bockman, M.W. *A Russian Space Station: The Mir Complex.* NASA-JSC.

Brinkley *et al.* The *International Space Station: An Overview.* IAF-97.

Butler, G.V. *Working in Space.* AIAA.

Caprara, G. *Il Libro dei Voli Spaziali.* Vallardi.

Cline & Gibbs. *Renegotiation of the International Space Station.*

Compton & Benson. *Living and Working in Space: A History of Skylab.* NASA SP-4208.

Cornett, L.H. *History of Rocketry and Astronautics.* AAS, vol. 15.

Covault, B. *Challenges, Promise Mark Station Debut.* AW&ST 12-8-97.

David, L. *Space Station Freedom: A Foothold for the Future.* NASA.

Dubbelaar, B. *The Salyut Project.* Progress Publishers.

ESA:

European Participation in the International Space Station. ESA MSM-PI/8041.

Utilisation of the International Space Station. ESA-MSM-4785.

Feustel & Buechl. *The European Participation in the International Space Station.* ESA MSM-CO/221-1998.

Freemann, M. *How We Got to the Moon.* 21st Century Associates.

Friedenthal, M.J. *Space Station Beyond IOC.* AAS, vol. 59.

Gartman, H. *L'Avventura Astronautica.* Bompiani.

Gearing, McKissock, Cookson, Ahlf & Huckins. *U.S./Russian Cooperation on the Space Station: The NASA/NPO Energia Studies Contract.* IAF-93.

Girain, G.A. *Mir-1 Space Station.* NPO Energia Ltd.

Glouchko, V.P. In *Encyclopédie Soviétique de l'Astronautique Mondiale.* Editions Mir.

Gore, A. "Remarks by the Vice-President in Signing Ceremony With Prime Minister Chernomyrdin of Russia." The White House.

Gray, J. *Space Manufacturing Facilities/Space Colonies.* AIAA-NASA.

Harland, D.M. *The Mir Space Station.* Wiley.

Hartford, J. *Korolev.* Wiley.

Hechler, K. *The Endless Space Frontier.* AAS History Series, vol. 4.

Heppenheimer, T.A. *Colonies in Space.* Warner Books.

Jacobs, D. & NASA. *International Space Station: Overview and Current Status.* IAF-95.

Johnson, Nicholas L. *Handbook of Soviet Manned Space Flight.* Science and Technology Series, vol. 48. San Diego: American Astronautical Society, 1980.

Kohrs, Huckins & NASA. *An Update on the Development of the Space Station/* IAF-93.

Kramer, S.B. *Early Engineering Design of Space Stations in the USA.* IAF-93.

Lardier, C. *L'Astronautique Sovietique.* Armand Colin.

Launius, R.D. *Key Documents in the History of Space Policy.* NASA History Office.

Launius, R.D. *NASA: A History of the U.S. Civil Space Program.* Krieger.

Launius & McCurdy. *Spaceflight and the Myth of Presidential Leadership.* University of Illinois Press.

Ley, W.L. *I razzi.* Bompiani.

Logsdon, J.M. *Exploring the Unknown.* NASA SP-4407.

Malyshev & Brant. *Integration of Different Engineering Cultures: A Challenge in the Space Station Program.* IAF-97.

Mark, H. *The Space Station: A Personal Journey.* Duke University Press.

McDougall, W.A. *The Heavens and the Earth: A Political History of the Space Age.* Basic Books.

Narita & Ogo. *JEM Operations Implementation Plan.* IAF-97.

NASA:

Alpha Station: Program Implementation Plan. Washington: NASA, September 1993.

Essays on the History of Rocketry and Astronautics. NASA Conference Publications 2014.

The First Lunar Landing, As Told by the Astronauts. NASA.

International Space Station: Familiarization. NASA-JSC.

International Space Station: Reference Guide. Boeing & NASA.

Our First Space Station. NASA, 1977.

Report to Congress on the Restructured Space Station Program. NASA, March 1991.

Space Settlements, A Design Study. NASA.

Space Station: A Research Laboratory in Space. NASA PAM-512.

Space Station Freedom Media Handbook. NASA, 1989, 1992.

Space Station Operations Task Force. NASA, 1987.

NASDA. *JEM: Japanese Experiment Module, ISS.*

Newkirk, E. *Almanac of Soviet Manned Space Flight.* Gulf.

Noordung, Hermann. *The Problem of Space Travel: The Rocket Motor.* NASA, 1995.

NPO Energia. *From First Satellite to Energia-Buran and Mir.*

Oberth, Hermann. *Uomini Nello Spazio: Come Raggiungere la Luna.* Milan: Loganesi, 1957.

O'Connor, B. *Final Presentation to Advisory Committee for Redesign of the Space Station.* NASA, June 1993.

O'Leary, B. *Project Space Station.* Stockpole Books.

O'Neill, Gerard K. *Colonie Umane Nello Spazio.* Milan: Arnoldo Mondadori, 1979.

Outlook for Space Study Group. *Outlook for Space: Report to the NASA Administrator.* 1976.

Parkinson & Smith. *High Road to the Moon.* British Interplanetary Society.

Portree, D.S.F. *MIR Hardware Heritage.* NASA-JSC 26770, 1994.

Reibaldi, G. *et al. The Microgravity Facilities for Columbus Programme.* ESA Bulletin, May 1997.

Robinson, D.W. *The Development of Space Station Objectives.* IAF-93.

Sagdeev, R.Z. *The Making of a Soviet Scientist.* Wiley.

Sogroske, V.P. *A Guide to NPO Energia Services.* NPO Energia Ltd.

Stuhlinger & Ordway. *Wernher von Braun.* Krieger.

Thirkettle, Gulpen & Boggiatto. *Highlights From the Development of the Columbus Orbital Facility.* IAF-97.

U.S. Government:

Civilian Space Station and the U.S. Future in Space. U.S. Congress-OTA-STI-241.

Salyut: Soviet Steps Toward Permanent Human Presence in Space. Congress of the United States-OTA.

Soviet Space Programs 1971-1980. U.S. Government P.O.

Vallerani, E. *Manned Space Station Systems from Alenia.* Alenia Aerospazio.

Vallerani, E. *L'Italia e lo Spazio: I Moduli Abitativi.* McGraw Hill.

Von Braun & Ordway. *Histoire Mondiale de L'Astronautique.* Larousse.

Winter, F.H. *Prelude to the Space Age.* National Air and Space Museum.

General References:

Agreement: 1993 to 1997. IAA-97/IAF-97.

Blueprint for Space: Science Fiction to Science Fact. Washington: Smithsonian Institution Press, 1992.

Cosmonautics: A Colorful History. Washington: Cosmos Books, 1994.

De Saliout à MIR. Moscow: Éditions Novosti, 1987.

Études Soviétiques. Paris: ES, April 1981.

The International Space Station: Engineering the Future. Status Summary, March 1994.

La "Saliut" en orbita. Moscow: Agencia de Prensa Novosti, 1985.

The Science Fiction Encyclopedia. New York: Doubleday, 1979.

The Space Station: An Idea Whose Time Has Come. New York: IEEE Press, 1984.

Space Station Development Status: Alenia's Participation in ISS. IAF-97.

Tant qu 'à faire, mieux vaut être le premier. Moscow: Éditions Novosti, 1991.